D1711552

DEMOCRATIC DILEMMAS IN

THE AGE OF ECOLOGY

Democratic Dilemmas in the Age of Ecology

TREES AND TOXICS IN THE AMERICAN WEST

Daniel Press

DUKE UNIVERSITY PRESS Durham and London 1994

WITHDRAWN
J. ROBERT VAN PELT LIBRARY
MICHIGAN TECHNOLOGICAL UNIVERSITY
HOUGHTON, MICHIGAN

TO DAN MAZMANIAN AND

ALISON RUSSELL

© 1994 Duke University Press
All rights reserved
Printed in the United States of America on acid-free paper ∞
Designed by Cherie Holma Westmoreland
Typeset in Palatino with Frutiger display by Tseng Information Systems, Inc.
Library of Congress Cataloging-in-Publication Data appear on the last printed page of this book.

GE
185
.W47
P74
1994

Contents

MAR 29 1996

Acknowledgments

Many people have helped me reach for, and write, this book. Thanks to Dan Mazmanian and Monty Hempel for encouraging me to study interesting things, and for prompting many of the best ideas in this project. Thanks also to Al Louch, who spent more time with me one-on-one than any professor should have to, and always with patience and graciousness. Tom Rochon, Porto, and Michael Lippitz have each been an infinite source of advice—both strategic and academic—and friendship.

For assistance with my field research, Doug Leedy was My Man in Oregon, a true spirit of Nature. Mark Sigel also helped enormously with things Oregonian. A few people extended themselves and secured new respondents for me beyond my best expectations. They are Faye Beuby, Gabriela Goldfarb, Catherine Morrison, Ted Trzyna, and Jerry Moles. Of course I want to pay particular thanks to the many respondents who gave so much of their time, each with his or her own angle of repose.

Research takes time, time is money, and I would have had none of these if it hadn't been for a generous dissertation grant from the John Randolph and Dora Haynes Foundation. The entire staff at EPICS International displayed great forbearance in letting me work part time but treating me like a full-time colleague. I'll never be able to thank David Morell, the president of EPICS and now managing associate at ERM-West, enough; he provided advice, compassion, support, and above all, approval when they were most needed.

A number of people helped me obtain important articles and commented on or clarified theoretical and methodological issues with me. They are Karen Hult, Russell Dalton, John Pierce, Riley Dunlap, Paul Sabatier, and Fremont Lyden. Several courageous friends listened to presentations of the research in various stages and gave helpful comments: Ted Glasser, Barbara Bair, Andrea Asaro, Michael Stanley-Jones, and Lynne Trulio. I also had generous help from staff in the Center for Politics and Policy, notably Sandra Seymour and Gwen Williams. Thanks also to my editors at Duke University Press, Larry Malley and Reynolds Smith, and to the two anonymous reviewers who were so helpful.

My family helped me do what needed to be done through it all. Bill and Jean Russell provided a haven in Claremont as well as great solicitude. Talya Press-Dahan, Esther McManus, and Richard Press gave me unconditional encouragement and inspiration. Of course, my wife, Alison Russell, takes the cake for patience, skillful editing, and love.

CHAPTER ONE Environmentalism Returns to the Democratic Fold

I think human beings are capable of doing very stupid, very violent things, we can see that in any ethnic group, in any culture; we can see some awful things that mobs and individuals do. And what frightens me is that, as the ecological crisis becomes more severe, as the heat is turned up, as the pie we're all trying to split up becomes smaller, and more people are trying to get it, as people really begin to see that resources are not infinite, people are going to get very, very weird.
—Dave Foreman

We are now in the Age of Ecology. Environmentalists exhort and pressure us to incorporate their concerns into nearly every major policy issue and arena, and policymakers are compelled to listen. Specific positions are bitterly contested as environmental lobbyists and their opponents swell the ranks of the Washington-based special interests. Thirty years ago, we might have guessed that environmentalism would simply become more raw material for the American machine of liberal democracy. Distributive or redistributive politics would more or less take care of the environmental problem in the usual ways.

But the Age of Ecology confronts us with new challenges and dilemmas. Environmental critiques of modern society go to the very heart of our political and economic organizations. Never before have we been confronted with such intractable problems, with threats that arise from within our own practices, and with such profound uncertainties over how to proceed. Environmental protection and restoration are not technically overwhelming—we probably had less of the requisite know-how for putting a craft on the moon in the 1950s than we do for solving major environmental problems today. In our society, environmental problems are *democratic* dilemmas. The Age of Ecology requires us to build seemingly impossible bridges: somehow hundreds of millions of individual actors must learn the ecological consequences of their behavior, and somehow they must use their knowledge to choose wisely between painful trade-offs. We must bridge the gap between local control over unwanted land uses and state or national interests, weigh the long-term costs of resource depletion and degradation against short-term economic dislocations. All this must be achieved in a society that has grown so large and contentious as to be effectively gridlocked and often incapable of widespread, meaningful deliberation.

We have been warned of these challenges for at least the last 25 years. Economists, scientists, and other environmental analysts have written that society may be in store for a rude awakening if we continue to address large-scale environmental problems with our cautious, incremental style of politics and policy. Our predicament, they all seem to agree, is that we move to action too slowly, we know too little, and our environmental policies are hopeless palliatives. And even when we think we know what to do, good policies are blocked by some interest or another.

A "centralist" group of theorists writing about these dilemmas argues that we have one of two grim choices: we can continue business as usual and try to adapt to life in an impoverished environment, or we can abdicate democracy in favor of an all-powerful but ecologically enlightened Leviathan.[1] "Decentralist" writers, on the other hand, urge us to devolve political power to small communities, where all could "think globally and act locally" and thereby preserve—even restore—democracy while saving the physical environment.[2]

Thus, the appropriate balance between democracy and environmental protection is continually being sought; the questions it raises are not merely theoretical but in fact are being asked self-consciously by people struggling for political goals in environmental controversies across the United States. Examples of environmental conflicts that raise these questions as central features of their debates abound. William Boyer, a political reformer in Oregon, wanted to write into the state constitution basic "environmental rights." He proposed to create these democratically, through a 1992 proposition that, however, never got onto the state ballot. The proposition was ambitiously called "the Oregon Plan for Environmental Rights," and would have created a five-person elected "Commission for Environmental Enhancement and Protection" that would consolidate control over the state Environmental Quality Commission, the Energy Facility Siting Council, the Water Resources Commission, the Board of Forestry, and the Fish and Wildlife Commission. All the elements of the theoretical debate mentioned above are present: a move to centralize control, albeit through the electoral process, an effort to guarantee certain environmental outcomes, and a palpable frustration with technocratic inertia and its characteristic inability to coordinate natural resources management.

Clearly, Boyer's initiative defined social justice in environmental and procedural terms. He wanted not only to codify and protect "environmental rights" but also to flex the muscles of the electorate. Boyer believed that environmental rights can be voted into the state's constitution and trusted that, given the ability and opportunity to do so, Oregonians would insert into it important environmental values. Californians have also sought to adopt environmental initiatives in earnest; Proposition 65, "Big Green" (Proposition 128 of 1990), and many other policy initiatives have meshed grass-roots participatory rhetoric with environmentalism.

The well-known NIMBY (Not-In-My-Back-Yard) syndrome exemplifies the struggle between local and centralized forces to manage environmental problems, especially in the hazardous waste arena. Centralists and decentralists agree that NIMBYism is a manifestation of both political frustration and the assumption that local control ("power to the people") can solve—or at least avoid—environmental dangers. Many community residents who are in the vanguard of NIMBYism assume that they would not have let toxics get so bad if they had played a role in the political process as equals with industry and government.

The 1986 "Tanner planning" process, named for AB 2948,[3] a bill sponsored by California Assemblywoman Sally Tanner, attempted to address NIMBYism at both the county and state levels. The Tanner bill required all counties in California to indicate in a plan submitted to the state Department of Health Services[4] how much hazardous waste treatment capacity that county would need to treat hazardous wastes in the future. Recognizing the volatility of facility siting conflicts, the bill also required counties to develop criteria for approving such facilities, and created a process for brokering agreements between facility sponsors, counties, local residents, and the state. Tanner planning has been perceived alternatively as a landmark in local, collaborative planning and as a vehicle for state preemption. Successful or not, the Tanner bill addresses issues of inclusion, jurisdiction, and centralized versus decentralized control over hazardous waste management.

Another toxics example displays yet again the battle over the appropriate locus of control for environmental management, this time in the San Gabriel Water Basin. This important source of groundwater in southern California became contaminated with toxics from the BKK Superfund site in West Covina and hundreds of industrial facilities. Local frustration with the slow pace at which the U.S. Environmental Protection Agency (EPA) was handling the San Gabriel Basin case prompted Congressman Esteban Torres to circumvent Superfund cleanup of the groundwater basin on the assumption that a local consortium could do the job better, faster, and with more consensus than the EPA. Whether local attempts to clean up the San Gabriel groundwater will be successful remains to be seen, but the interesting point is that local players are trying to bypass the large, powerful, centralized bureaucracy (EPA) precisely because of its inefficiency and undermanagement.

Northern California provides an excellent example of these topics in its attempt to reform (or, alternatively, to maintain the status quo on) timber practices. In 1990, a coalition of environmental groups and timber companies—the most visible members being the Sierra Club and Sierra Pacific Industries—hammered out an agreement subsequently dubbed the "Sierra accords." In light of other, unsuccessful environmentalist initiatives (especially the "Redwood Summer" protests and the "Forests Forever" and "Big Green" ballot initiatives of 1990), further curtailment of harvesting on behalf of the spotted owl's habitat, and mounting tensions in rural communities, these players were eager to avoid traditional arenas of conflict. The results of several months of collaborative policy meetings were four bills (later consolidated into one) submitted to Governor Pete Wilson for approval. He vetoed the legislation, claiming that the measure was too restrictive, hadn't considered his proposed amendments (he wanted the bill to incorporate over 80 changes), and, most important, had inappropriately attempted to treat the coastal forests and Sierra (inland) forests with the same rules.

Indeed, most of the coastal environmentalists and timber industries objected to the legislation, demonstrating both the fragility of participatory policy-making processes and the difficulties inherent in deciding appropriate jurisdictions for managing environmental dilemmas. In the end, the governor's own bill failed to pass a bitter and resentful legislature, and several rural northern counties reopened an old proposal to secede from the rest of the state!

While the legislature and governor wrestled with their misnamed "accords," local antagonists in California's north coast counties were trying to build understanding and consensus on forest management and watershed protection. Calling themselves "bioregional councils" in the Klamath Province, a loose network of advisory groups, residents, and traditional enemies began to seek—and sometimes received—authority over local natural resource decisions.

In large urban areas, public frustration with complex, interrelated problems such as transportation, air and water quality, park management, wetlands protection, and affordable housing has given support to new calls for regional government in urban areas. In northern California groups with names like Bay Vision 2020, the Bay Area Economic Forum, and Bioregionalists, have all urged the state legislature to consolidate transit districts, air and water quality control agencies, and numerous other functions of state and local governments. As of spring 1994, one major "regionalist" bill was working its way through the California state legislature. The new environmental superagency Cal-EPA has also called for regional management of its environmental review and permitting processes. Not surprisingly, no one is sure just how regionalism could help or hurt, and many municipalities are afraid of losing autonomy over the few issues that still remain in their jurisdiction. Above all, there is no consensus on what the appropriate locus of control for a particular problem might be.

Even the marketplace addresses some of these themes, in the form of "green consumerism," the ultimate in decentralized, individual environmental action. With companies like Working Assets offering telephone lobbying services along with "green investment," we see a move toward political empowerment pursued almost as much for its own sake as for the causes it serves. And in typologies of environmentalists, progressive companies like Ben & Jerry's Ice Cream or Working Assets don't seem to fit.[5] These companies, and many others like them, attempt to make good on the potential of "voting with your dollars" in a way that maximizes personal involvement and successful environmental outcomes.

The examples cited all share some common democratic dilemmas: Who should rule? On what basis? What are the environmental implications of centralized rule or decentralized rule? Can citizen responsibility foster sufficient competence for people to make decisions that are in the public interest?

Whenever they address these questions, centralists and decentralists carry on their debate at an abstract, theoretical level, but their analyses are lacking in two important respects: First, almost no empirical work has been done to substantiate theoretical claims about the relationship of environmentalism to democracy. This is true largely because there has been hardly any dialogue between theorists writing on environmental politics and those focusing on the difficulties of maintaining democratic practices in large, complex nation-states. As a consequence, environmental political thought has been influenced more by economics, biology, and ecology than by political theory. This environmentalist political work tends to be long on ecological critiques and polemic and short on understanding of the political process. Democratic theorists, for their part, have missed an opportunity to place democratic dilemmas in a context where they are current and compelling. The merger of environmental political thought and democratic theory can yield testable propositions for the claims of these two respective literatures.

Second, the explanations of why democracy and environmentalism should be logically incompatible are few and incomplete. Most claims of such an incompatibility do not define the concepts that are purported to be in conflict. "Democracy" is alternatively representative democracy, interest group liberalism, peer democracy, participatory democracy, and so on. "Environmentalism" is sometimes a profound critique of nearly everything associated with industrialized economies (see Kassiola, 1990), and at other times a justification for adopting standard operating procedures (like writing environmental impact reports) easily accommodated by the existing administrative state.

Failures of "American-Style" Democracy: Six Hypotheses

A number of testable propositions or implications emerge from the centralist-decentralist debate. They are claims about the proper locus of authority, the value and need for institutional limits, the ability of citizens to manage and understand technology, and the trade-offs between economics and environmental protection. Six hypotheses emerge from centralist and decentralist claims:

1. *Crisis Leadership:* If people perceive that they are in an environmental crisis, they will seek centralized leadership.
2. *Local Control equals Better Outcomes:* People perceive that local control over environmental management results in more desirable environmental outcomes than central control (e.g., state and federal).

3. *Outcome Determines Legitimacy:* People who participate in the policy process over environmental dilemmas judge the value of their experience by what happens in the physical environment.
4. *Participation Yields Control:* People who participate intensively on environmental issues gain more control over environmental policy outcomes.
5. *Technical Complexity Limits Participation:* Technical complexity is used by experts to block or negate public participation on environmental issues. Similarly, non-experts affected by environmental controversies limit their own participation, because technical complexity intimidates them, and prevents them from identifying, and effectively advancing their goals.
6. *Economics Limit Deliberation:* Concerns over economic growth sharply constrain the scope and number of distinct environmental policy options that are seriously considered by participants in the environmental policy process.

This study addresses the empirical gap in environmental political theory by testing the hypotheses listed above through in-depth interviews with respondents in two environmental controversies—forest management and hazardous waste management. Testing the hypotheses involves two tasks: first, to clarify which elements of democracy (both the liberal and participatory traditions) and environmentalism appear to be in conflict. Pared down to their essential differences, three reasons emerge to explain why theorists have claimed that democracy conflicts with environmentalism. I call these the "challenges" to democracy: The challenge of social justice raises the possibility that environmental protection may be a just course for society even if it is taken undemocratically. The challenge of technocracy strains our ability to bridge the goal of participation with the perceived need for expertise. And the economic challenge consists of the constraints that economic activity places on political decisions.

A second task is to show how far theorists' analyses can go in explaining successes and failures (in effect, the limits) of democratic politics in two current environmental controversies. Testing theoretical claims requires asking how environmental problems pose a challenge for American democracy and how institutions of democratic governance are responding. Note that what is considered the dominant form of political organization is American-style interest group liberalism. It is further assumed that this form of political democracy will not change dramatically in the near future, but that there may be opportunities to transcend its shortcomings. A number of the theorists described in chapter 2 have argued for a radical change in political forms; while the findings reported here may contribute to such claims, the main purpose of this study is to show how existing structures fail and to explore the emergence of new environmental politics.

Using respondents' policy preferences to assess the future of democratic environmental politics is a deliberate methodological choice. Most theorists evaluate democratic political forms on structural grounds and thus confine themselves to discussions of the *logical* effects of adopting certain procedures, decision and inclusion rules, and safeguards against abuses of power. But writers like William Ophuls and David Orr also stress that attidunial and value changes ("metanoia") must precede the transition to environmentally successful political structures. Thus, a critical step along the way to an ecologically sustainable polity is the transformation of citizens' political habits and attitudes toward the environment.

What evaluative criteria can be used to determine if American-style democracy is undergoing such a transformation? Given the plethora of environmentalist categories and democratic forms, *the most stringent evaluative criteria are those that ask how existing political democracy advances the core goals of environmentalism and democracy.* In the case of environmentalism, the core goal is to manage the biosphere in a sustainable manner. In the case of democracy, the core goal is to ensure that citizens can have meaningful participation in their society's choices. Therefore, American-style democracy will be evaluated based on its ability to conceive and implement sustainable environmental policies through highly participatory forms of governance.

For example, are structures emerging that attempt to accommodate the demands of participants, the ethics of environmentalists, the science of ecologists, and the underlying imperative of economic growth? If they are emerging, how have these structures made liberal democracy, industrialism, or the administrative state susceptible to environmental prerogatives, varied as these may be?

If, on the other hand, participants in environmental disputes are finding it difficult to maximize both democratic discourse and environmental action, what does the empirical evidence suggest is the cause of this failure? Theorists have asserted that the problem lies in everything from capitalism, liberal democracy, the administrative state, and industrialization to a cultural preoccupation with life and problems in the present or very near future. Whether such claims are accurate has been difficult to verify, due to the absence of systematic assessments of environmental policies' effectiveness in terms of the hypotheses listed above.

The Empirical Study

The field research for this study consisted of interviews with over 50 respondents in two environmental controversies, those surrounding forest management in Oregon and California and hazardous waste management in California. Respondents were asked questions that would help infer their views on the validity of the six hypotheses listed above (these are described in detail in chapters 5 and 6). The main findings of the empirical study follow:

1. Environmentalists are pragmatic, they seek local control of environmental policy when they can best show how local communities suffer from environmental degradation. Thus, where public health and "social justice" were equated with environmental protection, environmentalists sought political decentralization.
2. People often reacted to environmental crisis by increasing their participation, not by demanding more centralized control.
3. Respondents mobilized more to prevent losses than to make gains and thus exerted "negative authority" over environmental policy.

These conclusions should lead to a revision in our thinking about the issues that emerge from the centralist-decentralist debate. They also suggest that participants in environmental cases can learn a great deal about both technical matters affecting environmental issues and ways of influencing the political process. The findings of this study also make clear that high levels of participation may be expected in many environmental cases, but that coordinating such participation and reaching consensus on policy options can frustrate people's democratic habits. Many people perceive that their own environmental crisis is already upon them and have not hesitated to act.

Public perception of crisis is important because it serves as the link between what some theorists, especially centralists, describe as the inevitable course of industrialization and a testable proposition of theoretical claims. Clearly, the arguments of centralists and decentralists cannot be proven or disproven if we assume that no one can know how people will react to environmental crisis until environmental problems reach catastrophic proportions. If instead the important question is "Do people abandon democracy when they *perceive* themselves to be in an environmental crisis?" then researchers can begin to understand the dynamic between environmentalism and democracy before people feel the effects of long-term global problems.

Chapter 2 begins by revisiting recent "environmental political theory," stressing that the terms "environment" and "democracy" are used so vaguely that researchers may as well have been arguing over semantics and typologies as over substantive issues in political theory. Chapter 2 gives an overview of these substantive topics and explores potential incompatibilities between some forms of environmentalism and different kinds of democracy. Chapter 2 goes on to explore how various approaches to mitigating the shortcomings of democratic governance may be used to transform the policy process in favor of environmentally sustainable political choices. In chapter 3 the three challenges to democracy are explored: First, if environmental problems get bad enough, environmental justice might be ranked above the "niceties" of democratic principles. This is the "social justice challenge to democracy." Second, the complexity of environmental issues might require that technocracies marshal the forces of science in the cause of environmental protection, without hindrance from uninformed citizens. This is the "technocratic challenge." Finally, the primacy of quantitative, extractive economic growth might have foreclosed many environmental policy options, even to the point where democratic policymakers cannot afford to consider them without being punished by actual—or potential—downturns in the market. This is the "economic challenge to democracy."

Chapter 4 shows what we have learned from large-scale survey data and lays out the in-depth interview methodology used in this study. This chapter identifies questions that could not be asked in large surveys, and demonstrates how the six hypotheses listed above are explored by means of an intensive interview format.

The empirical research discussed in chapters 5 and 6 consists of in-depth interviews with respondents in two environmental conflicts. The hazardous waste management case exhibits classic examples of NIMBYism and public concern over human health risks. The forest management case demonstrates conflicts arising over alternative land-use visions—rather than disputes over health risks—for the oldest forests remaining in the United States. The two cases pose different problems for democratic theory; respondents' views on participation, information, and successful outcomes are compared in the two cases, and trends are reported. Chapter 6 examines again the three challenges to democracy to see if they played a big role in hampering the capabilities of participatory democracy to effectively address these environmental problems with sustainable, fair environmental policies. Chapter 7 concludes the study by examining some of the implications of the earlier chapters for the design of institutions to simultaneously uphold sustainable environmental policies and participatory democracy. This final chapter closes with a reassessment of the theoretical positions staked out by the main proponents of centralization and decentralization.

CHAPTER TWO Environmental Political Thought and Democratic Theory

Just as an awareness of environmental problems gathered momentum in the 1960s and 1970s, political theorists, economists, and biologists began to suggest that the pressures of the coming environmental crisis would become so strong that they would destroy democratic political institutions and the economic systems they relied upon. The implication was that no mode of democratic politics currently practiced could possibly handle problems of the scope and complexity of global environmental degradation. Democratic politics would be too slow to respond, relying as it did on incrementalism and drawn-out public policy-making. Modern democracies were also faulted for being incapable of protecting the interests of future generations.

Environmentalist writers began to claim that if democracies couldn't marshal the consensus and foresight to save the physical environment, maybe centralized power would cut through delays and small-minded policy-making. Enlightened authoritarian regimes could merge ecological awareness with swift, forceful, and comprehensive action to restore ecosystems and restrain human appetites for unsustainable growth. But the alternatives to democracy did not look promising to writers committed to democratic participation. Why should an authoritarian state be a priori better able to cope with environmental problems? After all, centralized bureaucracies are not known for their flexibility, responsiveness, adaptability, or forward-thinking capabilities.

Thus, very early on in the 1960s and 1970s debate modern societies were apparently stuck with a grim choice: either they had to give up the hope of altering their activities fast enough to avert catastrophe, or they had to forego their cherished democratic ideals in favor of a safer physical environment (albeit an uncertain social future).[1] The writers who articulated this "centralist" analysis have been called "neo-Hobbesians,"[2] "neo-Malthusians,"[3] and "structural reformers"[4]; they include Robert Heilbroner, Garrett Hardin, and to some extent, William Ophuls.

The ecological predicament described by these centralists was destined to set the tone of environmental political theory for the 1970s. In 1968 Garrett Hardin published his influential piece "The Tragedy of the Commons," which presaged environmental crisis unless society agreed to self-restraint through mutual coercion. Hardin's "lifeboat ethics" made the analogy of the earth as an ocean-bound vessel with limited space and limited food to feed its crew. Survival in the lifeboat would require strict discipline and the imposition of rules—for both using resources and giving them away. An ever-increasing number of would-be passengers had to be refused a place on board if the vessel was to avoid sinking. Indeed, Hardin, a biologist, found it to be a greater disservice to prolong the existence of the starving poor than to let them die out, dwindling in number until their populations were sustainable. He also asserted that such an outlook was likely to be considered draconian, perhaps even immoral, and thus only very strong social control could avert the tragedy of the commons.

Economist Robert Heilbroner followed in 1974 with his *Inquiry into the Human Prospect*. Heilbroner agreed with Hardin's analysis, adding that "the pressure of political movements in times of war, civil commotion, or general anxiety pushes in the direction of authority, not away from it."[5] For Heilbroner, such a tendency exists because people have a strong inclination toward obedience, especially of parental figures whose authority is deemed "up to the task" at hand, and also because a sense of social belonging gives people a large capacity for national identification. In turn, such nationalist identification could render society more susceptible to demagoguery. Heilbroner's analysis of political and economic institutions led him to believe that environmental crisis would surely accelerate so fast that societies would be compelled to impose ecologically rational behavior, using authoritarian means if necessary. But Heilbroner was not just projecting what he thought the future of politics would hold in a world saved from environmental disaster; he was willing to curtail democracy if by doing so environmental crisis could be averted: "I not only predict but I prescribe a centralization of power as the only means by which our threatened and dangerous civilization can make way for its successor."[6]

As for decentralized models of environmental governance, Heilbroner argued that "mankind lives in immense urban complexes and these must be sustained and provisioned for a long time. . . . pockets of small-scale communities may be established, but they will be parasites to, not genuine alternatives for, the centralized regime that will be struggling to redesign society."[7]

Shortly after Heilbroner's frankly authoritarian statement, William Ophuls published *Ecology and the Politics of Scarcity* (1977). In what is still one of the most widely read treatises on environmental political theory, Ophuls worked his way from a description of the tenets of ecology, through a catalogue of environmental ills, to a critique of laissez-faire economics and status quo politics. His intention was to demonstrate that the logic of healthy, functional ecological systems could only work within limits. Growth beyond such limits would constitute, in effect, living off one's capital (at increasing rates) instead of one's income.

Ophuls then contrasted the ecological limits to growth model of the world with the wasteful, consumption-driven model of the dominant economic paradigm. He faulted neoclassical economics primarily because it had supplanted politics[8] by "substituting growth for political principle." Economics thus "solves" ecological dilemmas through economic bargains even though they "are not easily compromisable or commensurable, least of all in terms of money."[9]

If political principles were gone, replaced by an economic ideology devoted to infinite growth, Ophuls was convinced that only sharp, coercive actions would allow any kind of survival. He predicted that if society adopted steady-state economics under a crisis regime,[10] the new system might be run by ecological mandarins who would possess the esoteric knowledge needed to run it well.[11] This ecological oligarchy would have to curtail personal freedoms in order to make and implement prudent decisions.

But Ophuls was not a centralist in his recommendations; on the contrary, he concludes his book with a chapter devoted to the importance of participatory democracy, value change, and enlightenment. Thus in his prescriptions Ophuls is very much a decentralist, a point he emphasizes even more strongly in the 1992 second edition of his book. Calling Hardin's solution to the tragedy of the commons "explicitly Hobbesian," Ophuls points out that it was only his *analysis* of the issues that might be termed Hobbesian, not his *prescriptions*.

Most decentralists criticized Ophuls, Hardin, and Heilbroner for *both* their analyses and prescriptions. Decentralists found their analyses flawed, countering with centralization as the root cause of environmental problems and thus reaching a prescription opposite to that of Hardin and Heilbroner (recommending decentralization and participation as the basis for communicative and ecological rationality).

Through his 1974 book *Man's Responsibility for Nature* Passmore was one of the first to condemn centralized approaches. He stressed that democratic societies are in fact more capable of action than totalitarian states, especially because of their "habits of local action" and their "traditions of public disclosure."[12] Contemporary democracies thus fail "when they are tempted into exactly the same vices as the totalitarian states; spying, soothing utterances from central authorities, face-saving, bureaucratic inertia, censorship, concealment."[13]

David Orr and Stuart Hill (1978) offered a somewhat more precise critique of the centralists and questioned, in particular, their assumptions "(1) that an authoritarian state can cope with its own increased size and complexity; (2) that it can muster sufficient skill to exert control over the external environment; and (3) that these conditions can be maintained in perpetuity" (p. 459).

To these assumptions, they reply that large, highly centralized systems such as bureaucratic organizations tend to be limited in their ability to process complex information with a comprehensive understanding of the universe in which they must act. Orr and Hill also argue that government regulation and bureaucratic fiat have been poor tools for systemic approaches to ecosystem repair, and that truly creative problem solving tends to occur less in authoritarian groups than in their democratic counterparts.

Susan Leeson (1979) agrees that the authoritarian prescription is too simple and doesn't explain how an authoritarian system would operate "ecologically." She also points out the lack of proof we have for the ability—or motivation—of authoritarian systems to manage environmental problems better. Orr and Hill add that since resilience and learning are not characteristics of highly centralized systems, an ecological Leviathan whose legitimacy was predicated on successful environmental management would not have much margin for error. Moreover, successful environmental outcomes are almost always jeopardized by the "megaprojects" many centralized regimes have been so fond of in the past. Such projects would never be attempted if communities were small and self-sufficient (e.g., no larger than 5,000 people).[14]

The decentralists ultimately take issue with the centralists' "model of self-centered man who is incapable of acting in the larger community interest and incapable of deciphering the complexities of ecological problems."[15] On the contrary, "selective decentralization" would permit citizens to retain a high level of involvement in economies that were always "local." Democratic participation would not only be feasible as a political system and effective at safeguarding the environment, it might very well allow *ever increasing* levels of democratization beyond that which current societies had to offer.[16]

In a recent work, Orr (1992) reiterated his belief that decentralized governance is a necessary condition of achieving environmental sustainability. However, like most of the decentralists, Orr does not offer a convincing link between decentralization and sustainability. Why should decentralized governance lead to environmental health and sustainability? Aren't people in small communities just as capable of ransacking their resources and condemning land and water for waste disposal? In response, Orr would say that "the transfer of power, authority, resources, talent, and capital from the countryside, towns, neighborhoods, and communities to the city, corporations, and national government has undermined in varying degrees responsibility, care, thrift, and social cohesion—qualities essential to sustainability" (p. 71).

Orr jumps from civic virtue and participation to sustainability and does not fill the gap with testable propositions. Somehow, sustainability and environmental conservation arise from "good communities . . . and good livelihood."[17] The sentiment is appealing, but the connections between these desirable social traits and environmental action are not clear. Does social cohesion change a community's long-term discount rate, thereby encouraging it to value the future more? Is self-reliance somehow linked to greater participation and consensus building in natural resource management?

Perhaps the decentralists don't elaborate on these connections because they expect that structural (political, institutional) decentralization would have to *simultaneously* be accompanied by self-transformation. John Rodman (1980) stresses that the debate over structural choices reflects a mistaken assumption about how resource limits are imposed. If limits are external, certainly an authoritarian state may be the most direct way to impose them. If resource limits arise from our own dissatisfaction with a pointless life of production and consumption, "then . . . limits to [industrial] growth" emerge "naturally."[18] The implication for Rodman is that the real locus of change will not be in governments, democratic or authoritarian, but in a "new paradigm . . . [that] is more apt to be discovered than legislated."[19]

Rodman brings the link between decentralization and sustainability into focus, elaborating on Orr and Hill: he associates totalitarianism with ideological monocultural purity ("new Aryan order," "the new Soviet man"), and points out that environmentalists' preoccupation with biodiversity is based on the same ecological reasoning that concludes that totalitarianism is unsustainable; in essence, uniformity breeds instability. This is what Orr means when he says that authoritarian systems are unstable and that "ecosystems are the only systems capable of stability in a world governed by the laws of thermodynamics."[20] Thus, for Rodman, the struggle for preserving biodiversity is "a resistance movement against the imperialism of human monoculture."[21]

Resistance to the inertia of the status quo and the logic of existing economic and political paradigms is the key to the decentralist position. Decentralists reject the authoritarian trade-off between democratic participation and desirable environmental outcomes, thereby simply refusing to rank environmentalism above democracy, or vice versa. In a prescient close to his 1974 book, Passmore summed up this desire to simultaneously maximize good environmental outcomes and preserve just social institutions: "My sole concern is that we should do nothing which will reduce their [future generations'] freedom of thought and action, whether by destroying the natural world which makes that freedom possible or the social traditions which permit and encourage it."[22]

Paradigm Change through Transformative Politics

Many of the writers discussed above argue that, whatever political path we choose, radical paradigm changes will be necessary to avert environmental crises. Eckersley (1992) points out that much environmental political thought views environmental crisis as an "opportunity for emancipation." She suggests that this emancipatory potential has motivated ecopolitical theorists to direct "considerable attention toward the revitalization of civil society rather than, or in addition to, the state."[23] Can this emancipation or transformation occur under Leviathan rule? Probably not without discarding basic premises of the centralist analysis. The centralist perspective sees a logical contradiction to giving people in the present control over the future. As individually rational actors, people are compelled to destroy collective goods for individual pursuits, and hence they are not even capable of the "metanoia" Ophuls urges. A Leviathan concept—even that of Ophuls's ecological mandarins—implies a regulative approach to mitigating environmental change instead of collective attitudinal transformation.

In contrast, decentralists argue for political decentralization not just for better environmental management but as the best mode for paradigm change. Decentralists might say that you can't "think globally" (with new paradigms) without "acting locally" (through participatory democracy). But they have not suggested *how* democracy and participation can help transform and educate enough people to achieve a paradigm change. This threshold question is the subject of the rest of this chapter.

Following the decentralist line of reasoning, intellectual transformation is facilitated by, and requires, a participatory process. Why? Because people cannot learn, challenge new information, or deliberate in a closed, nonparticipatory system. And in contrast to environmental political thought, democratic theory has explored innovations, both in social relations and state structures, designed to achieve citizen transformation through political participation. Borrowing from democratic theory, we can ask a series of questions about how different forms of participation might bring about the kinds of attitudinal transformations decentralists predict if their vision becomes reality: How can people learn about complex issues? How can people act (or participate) on these issues? What social goods can people act upon? When people take control of policy, how will they impose limits upon themselves?

Transformation through Deliberation and Learning

Starting with participatory democracy's possible educational influence, a number of theorists suggest that policy learning is acquired through political practice. For example, Fishkin (1991) suggests we hold "deliberative opinion polls" on a national scale, whereby a small group representing as many interests as possible would come together to debate policy issues. Participants would meet face-to-face for at least two weeks, during which time they would be able to deliberate and learn enough to make decisions or recommendations with enlightened understanding. Their deliberations would be open for all to see, and their recommendations would be used to set policy agendas. As a way of influencing the choice of presidential candidates, Fishkin claims that the deliberative opinion poll would increase political equality and deliberation without risking tyranny. The deliberative opinion poll is fundamentally still a mechanism of representative democracy and does not address the nature of the information that must be learned and shared by the members of participatory governance approaches.

Thus we have to turn to research on the policy process, like Sabatier's (1988), to see what *kinds* of information and knowledge appear to be necessary for participants to successfully (and meaningfully) negotiate policy disputes. Sabatier describes how participants engaged in political dialogue form their beliefs and negotiate them with others. He offers insights on how deliberation and policy-learning processes affect the dynamics of individual and institutional behavior, especially as people interact in "policy networks."[24] Such networks form stable "advocacy coalitions" built over a long time in each issue arena. Members of advocacy coalitions are highly motivated to learn about their policy areas and come to adopt hierarchies of beliefs about politics and society in general and specific issues in particular. They share deep "core beliefs" that are very difficult to change, as well as "shallow" core and secondary beliefs. Sabatier's approach sheds light on what might happen in governance networks where there is a high degree of controversy over deep core, shallower core, or secondary beliefs. Policy change, in his view, is strongly dependent on the degree of policy learning that can be achieved *among* advocacy coalitions: "Policy-oriented learning across belief systems is most likely when there is an intermediate level of informed conflict between the two."[25]

Policy change and compromise are further aided by "a forum which is (a) prestigious enough to force professionals from different coalitions to participate, and (b) dominated by professional norms" (p. 156). Sabatier hypothesizes two further refinements that can help identify positive policy-learning situations:

> Problems for which accepted quantitative performance indicators exist are more conducive to policy-oriented learning than those in which performance indicators are generally qualitative and quite subjective. . . . Problems involving natural systems are more conducive to policy-oriented learning than those involving purely social systems because in the former many of the critical variables are *not* themselves active strategists and because controlled experimentation is more feasible. (p. 156)

Environmental policy dilemmas are often focused on natural systems and can be characterized, in part, by quantitative performance indicators (how much toxin exists, how many acres have been logged). To the extent that quantitative indicators are accepted by coalition opponents, better information about natural systems should allow learning to evolve. But this may only mean that a given policy network may achieve greater understanding over time—not necessarily consensus on policy options.

Transformation through Large-Scale Participation

If policy learning and deliberation are to succeed in contemporary societies, they must be supplemented by some vehicle for coordinating participation on a large scale. What kinds of large-scale participation can achieve attitudinal transformation? Favorite tools of environmentalists working at the local and state levels have been initiatives and referenda, or "direct legislation." With a rhetoric that dates back at least to the progressive movement, environmentalists insist that initiatives can bypass gridlocked legislatures by returning power to the people. Threats of toxic pollution and irreversible environmental losses lend urgency to this strategy, which has recently paid off several times in California.[26]

Ideally, initiatives give citizens quasi-legislative powers, increase participation, and educate people about political processes. In the 1980s they were also used to alert voters to issues that could not wait for incremental legislative action, or to bring policies to a vote before they could be watered down by political horse trading.

In *Direct Legislation,* Magleby (1984) shows that in practice direct legislation often fails to deliver on its promise. The use of initiatives and referenda does not increase participation; instead, voters exhibit high "dropoff." That is, they do not vote on every proposition but just on the candidates (whose parties provide cues to voters) or on the first few initiatives on the ballot. Magleby found that voters were confused by complex proposition texts, misled by ballot titles, and sometimes led to vote in opposition to their true preferences. Most voters think the process could be improved.[27]

In general, voters on ballot propositions are unsure of their positions on obscure issues and are easily overwhelmed either by the sheer number of propositions or by savvy campaign appeals. The result is that representation is increased only for those who understand the initiative process and can use it to their advantage, except in cases like property tax reforms, where the message and the intent of the legislation are very simple and benefit large numbers of voters. The difficult language of propositions leaves an opening for powerful interest groups, who hire professional campaign consultants to get the required number of signatures, place the initiative on the ballot, and then launch a media blitz likely to sway voters in its favor.

Magleby also criticizes direct legislation for short-circuiting the normal process of accommodation and accountability that takes place in legislatures. The voters are presented with a finished piece of legislation that they can only vote up or down; they cannot take part in the drafting of bills and the compromises that should result in "better" legislation. Finally, there is no deliberative process and no way of assessing the intensity of preference expressed by the voters.[28] More important, initiatives and referenda—by default—are used as tools to shift responsibility and accountability away from elected officials on controversial issues.[29]

Magleby's critique notwithstanding, initiatives and referenda can be important tools of participatory transformation as long as they are supplemented by better understanding. In Fishkin's analysis, most initiatives increase values of political equality but at the expense of deliberation; thus, if deliberative and educative "correctives" could be applied to the initiative process, an old means of large-scale participation could be given new life. Campaign finance reform, better written (and brief) proposition texts, deliberative polls, interactive voting and dialogue systems,[30] and in some cases even functional representation[31] could all be used to answer Magleby's incisive criticisms.

Participatory Democracy and Attitudinal Transformation

For other theorists, the problems of democratic participation can be resolved not by tinkering with electoral or institutional fixes but by making democratic politics less adversarial. In *Beyond Adversary Democracy*, Jane Mansbridge explores this possibility. Mansbridge studied citizens participating in New England town meetings and at an inner-city crisis center, using lengthy personal interviews and social histories to arrive at the motivations of these small groups of "political participants."

Mansbridge's main purpose is to describe the prevalent form of democracy in the United States today—what she calls adversary democracy—

and then show it is not the only form of democracy practicable in a large nation-state. If adversary democracy is predicated upon constantly warring interests, the alternative form, unitary democracy, is distinguished by four basic characteristics: members have largely common interests, equal respect for each other, a strong desire to reach consensus, and an understanding that decisions should be worked out in face-to-face meetings.

One crucial difference between adversary and unitary democracy is that the former accepts outcomes that produce winners and losers, while the latter does not. In essence, they diverge on procedural questions, especially decision rules. Adversary democracies discount dialogue to the point of relying solely on "bargaining," that is, settling impasses by majority vote. In contrast, unitary democracies shun conflict, preferring to reach decisions by discussion and consensus, or by relying on unanimous vote (by secret ballot) only as a last resort. And voting is avoided because members of the unitary democracy are loathe to alienate people they have to live and work with each day. This is not to say that conflict in unitary democracies does not exist or that it is necessarily bad, but in some cases Mansbridge found that the fear of conflict may—in small groups—actually "goad" weaker members into acquiescence, thereby defeating the goals of dialogue.

The ideal of equal power for all (theoretically espoused in a unitary democracy) has always seemed highly impracticable, and thus most political scientists have labeled it a goal unlikely to be reached. But Mansbridge shows that the requirement of equal power for the democracy's members does not necessarily obtain in all circumstances. Equal power is not necessary to provide equal protection of interests if everyone has roughly common interests. Equal power is not necessary to ensure equal respect if respect can be derived from sources other than personal power. Finally, equal power may be sought after for reasons of personal growth or high participation in public affairs, but will only be a burden if a member has all the responsibility and conflict he or she can stand. Mansbridge says that egalitarians should "order situations along a spectrum ranging from complete identity of interest to complete conflict and should be willing to pay more to get equal power as interests diverge."

Some aspects of unitary democracy can be applied to a national scale, Mansbridge argues, notably:

1. Citizens can recognize the ideals underlying their democratic beliefs even as they acknowledge and sustain the practice of adversary democracy.
2. In the event of an external threat, citizens may believe, briefly, that there is a common good and that it counts above all else.
3. The national government can recognize the legitimacy of conflict a priori and get on with the business of resolving conflicts of interest in the aim of providing roughly equal protection to its citizens.
4. By viewing the national economy as highly interdependent, and by making it less competitive, the average worker would achieve a greater awareness of his or her common interest with others.

Each of these aspects requires some change in the ways people view democratic activity, as well as the political reach of their participation. Once again, this is a problem of political transformation, but in this case the question is whether there are some social issues or goods that lend themselves to less adversarial democratic methods of social choice.

In his article "Democracy and Self-Transformation," Mark Warren (1992) recently addressed this question by exploring the limits of an "expansive" democracy.[32] Expansive democracy relies on four premises: First, political interests and capacities are not fixed in some "prepolitical factors" but are determined also by institutions. Second, self-transformation is an important component of expansive democracy because it maximizes "opportunities for self-governance and self-development . . . not so much because it allows maximization of prepolitical wants or preferences."[33] Self-governing, in this view, is desirable not only because individuals can maximize their own gains but because "self-governance" is not simply a "private matter." Nonetheless, Warren stresses that self-governing and transformation through participation is essential to developing individual autonomy, because people must know what they want (they must be self-aware) in order to participate meaningfully within small- or large-scale democratic mechanisms.[34]

A third point is a familiar one: democracy has an intrinsic as well as instrumental value. Not only is democracy a means to achieve one's political goals, but it also serves as a way of learning how to articulate those goals and which matters to discuss in the public sphere. A fourth dimension to expansive democratic theory refutes some of the conundrums posited by earlier democratic theorists, namely that increased participation would make nations ungovernable. Warren argues that "increased participation is likely to encourage substantive changes in interests in the direction of commonality, transforming conflict in the direction of consensus."[35]

Not all social goods can be objects of decision in Warren's expansive democracy. It is the features of these goods that determine how successful participatory democracy can be in reaching acceptable resolutions. Key attributes of different goods are as follows:

1. Whether they are individual or social
2. Whether they are excludable or nonexcludable
3. Whether they are material (physical) or symbolic
4. Whether they are scarce or nonscarce

Warren gives names to each of the goods that exhibit different mixes of attributes and asks how participation, or "expansive democracy," would serve the acquisition and distribution of such goods.

Warren then poses three questions to determine (1) the level of conflict inherent in acquiring and holding the good, (2) whether the value of the good depends on social interaction and recognition, and (3) whether achieving the good requires common deliberation and action. Warren suggests that expansive democracy may be most suitable to nonscarce goods that depend on discursive values like social interaction, recognition, common deliberation, and common action.

Would expansive democracy lend itself to resolving environmental policy dilemmas like the forestry case examined in this study? To arrive at an answer, opponents have to agree on what the disputed goods are, both for analytical and prescriptive purposes. Specifically, are opponents concerned about *timber* or about *ancient trees?* If the disputes are over timber, forest goods might be classified as *individual material goods*. Timber is then scarce, individual, material, and excludable. That is, the resource is in short supply, it can only be consumed by individuals or small groups of individuals, and access to timber can be excluded (by cost and ability to use or distribute timber). In Warren's typology, individual material goods are not good candidates for expansive democratic structures. Such goods are inherently conflictual; moreover, their allocation has traditionally been handled by markets, not by deliberation and common action. Markets also determine the value of timber as a material good.

If forest goods are defined as ancient trees, they might be characterized as individual, symbolic, scarce, and nonexcludable. Ancient trees may be appreciated by individuals, and they are scarce. For those who define forest goods as ancient trees, forests may have symbolic attributes such as "otherness," nonhuman value, or "natural identity" (Maser, 1988; Oelschlaeger, 1991; Muir, 1970). Ancient trees as symbols help some people understand their own identities and responsibilities as humans. Arguably, ancient trees are nonexcludable since the fact of their preservation would seem to lend them common property attributes. But, of course, a tree left to stand can't be turned into lumber, and thus other definitions of the good (and therefore other users) are excluded. Although Warren doesn't describe the kind of good this might be one can imagine that the scarcity and exludability attributes might provoke conflict over forest goods as ancient trees.

Warren's analysis suggests that participants may have to redefine environmental benefits to bring them under the umbrella of expansive democracy. The definitional problem that participants face may hamper their attempts to move environmental goods from a "mostly individual" to "mostly social" status, and thus into a more favorable arena for their own participation and control.

Democratic Rules for Managing Resource Scarcity

Elinor Ostrom's work (1990) on common property resource (CPR) management complements Warren's focus on self-transformation. Ostrom emphasizes that the dynamics of individual behavior can be understood in light of changing institutional rules. In her inquiry, the transformation problem is one of choosing the right set of self-imposed rules for local common property resource management in the absence of coercion administered from outside the boundaries of particular CPRs. Ostrom catalogues three kinds of institutional rules that can affect individual users (or "appropriators") of CPRs:

1. Operational rules (e.g, agency permit decisions)
2. Collective choice rules (e.g., the statute governing an agency)
3. Constitutional rules, such as those governing legislatures (Ostrom, 1990; Sabatier, 1991)

Using a series of case studies, Ostrom shows how different appropriators in CPR disputes were able to develop new rules for managing and using their resources. Her work is important because she shows that CPR users in developed, urban areas, as well as in vastly different rural settings, have been able to change the institutions they operate under in similar ways despite cultural, political, and economic differences.[36] Ostrom argues that successful CPR institutions all exhibit a set of design principles, including:

1. Clearly defined boundaries
2. Congruence between appropriation and provision rules and local conditions
3. Collective-choice arrangements
4. Monitoring
5. Graduated sanctions
6. Conflict-resolution
7. Minimal recognition of rights to organize
8. Nested enterprises (for CPRs that are parts of larger systems)

Ostrom's analysis can be applied to many CPR cases but may have to be modified to fit issues like the forestry and toxics examples of this study. First, nested enterprises and jurisdictions very narrowly restrict the ability of appropriators to change governance rules. Thus, counties, loggers, and environmentalists in California's northern counties can come to agreement on management schemes for forests but may encounter roadblocks from state and federal agencies or policymakers. Second, in the forestry case, conflicts are not just over the level or rate of appropriation but also over whether there should be *any* appropriation at all. Activists challenge basic concepts of property rights in both public and private ownership. This distinction between Ostrom's cases and forestry or toxics suggests that difficulties will arise even at the constitutional rule level when participants question the underlying purpose of rule making (i.e., the standard reason is to place limits on appropriation, but environmentalists would demand that there be *no* appropriation in many cases).

Specifically, Ostrom's case examples do not treat CPR conversions. For example, the appropriation rules in the Turkish fishing rights, Swiss grazing, and Japanese wood-gathering cases of her study all assume that the common property resource's nature will remain stable. Its uses and natural features will not change significantly with sustainable use. But in the case of old-growth forests, environmentalists argue that timber harvesting and replanting really constitutes land conversion.

The toxics case provides a third and related difference with Ostrom's study. Many fears raised by hazardous wastes concern groundwater contamination but not necessarily groundwater *withdrawals.* Ostrom's approach addresses the quality of the CPR only through limiting the levels of appropiation. She does not explore the kinds of rules that could bring together self-interested parties to protect the quality of a resource, regardless of consumption levels. Rules for resource extraction (quantity) and resource protection (quality) might be the same as long as extraction is kept at palatable levels. When extraction rates exceed sustainable levels, the character of CPR rules may have to change, an aspect that Ostrom has not broached.

Conclusion

Whether it occurs through self-imposed institutional rules, policy learning, or deliberation, participatory democracy is almost synonymous with transformation. The very act of participation educates, if only by showing how narrow many of the avenues for political expression have become. New approaches can be explored, and if these demonstrate that large-scale voting mechanisms are meaningless, they may yet be transformed by deliberative means.

Participation requires that individuals identify and defend their beliefs. Theorists like Dryzek, Warren, Barber, and Mansbridge cannot be certain that a more subtle understanding of policy dilemmas will be sufficient to change attitudes and behavior. However, they can be sure that such an understanding is truly necessary to the task of paradigm change. But obstacles are likely to be thrown in the way of decentralizing policy issues as complex as those posed by environmental change. Despite the promise of democratic transformation, very potent contradictions and challenges to participatory models are hard to reconcile with American liberal democracy. These challenges are the subject of chapter 3.

CHAPTER THREE The Challenges to Democratic Environmental Policy: Social Justice, Technocracy, and Economics

Chapter 2 suggested that participation in environmental policy-making may be very problematic, but environmental political thinkers have not given a detailed analysis of why environmentalism and democracy, writ large, might be incompatible. For the purposes of this study, the strongest test of whether environment and democracy are incompatible would determine if policies that foster environmental sustainability are achievable by participatory democracy. By "sustainable environmental policies," I mean policies that *direct human activities in ways that do not irretrievably impoverish ecosystems or need-based development opportunities for future generations.* A policy perspective designed for maintaining a sustainable environment should thus accomplish the following:

1. Avoid irreversible action or alterations of the physical environment
2. Incorporate a consideration of lag times and delayed effects (similar to no. 1)
3. Sustain ecological diversity
4. Avoid displacement of problems through remediations or preventive action across time, media, and space[1]
5. Consider human components: cultural, economic, present and future generations

What is meant by "participatory democracy"? In essence, *participation at a level and a point where individual choices are meaningful.* Key characteristics:

1. Effective participation, including deliberation and political equality (Fishkin, 1991)
2. Voting equality at the decisive stage (where voting is used as a decision tool)
3. Enlightened understanding
4. Control of the agenda
5. Final control or decision (Dahl, 1989)
6. Legitimate decision-making process
7. Consensus
8. High degree of communication between participants (Dryzek, 1990)

9. Community-building; public interest, common goods, and active citizens (Barber, 1984)

This concept of sustainable environmental policies incorporates many of the assumptions held about environmental protection by both the centralists and decentralists.[2] The disasters they warn of are large scale, irreversible, and would impoverish or radically simplify ecosystems throughout the world. One may wonder at the quality of political life they assumed would be possible under an Eco-Leviathan, but the centralists certainly envisioned a physical environment restored to a level where nonhuman nature could still exist without a constant danger of breaking the earth's carrying capacity. If anything, the decentralists and steady-state economists assume that their vision of rational environmental politics offers a richer, more meaningful life, as well as a more ecological way to evaluate one's standard of living.

Focusing on participatory democracy gets to the heart of the centralist-decentralist debate, because it is over the role and potential of participation that they disagree the most. And participatory democracy is a recurring motif of American politics, what Morone (1991) has called "the democratic wish:" "The democratic wish imagines a single, united people, bound together by a consensus over the public good which is discerned through direct citizen participation in community settings" (p. 7).

The disagreement about participation takes place over opposing concepts of human nature, of the problem-solving capabilities of citizens, of the appropriate scope of policies (e.g., the extent to which governments can intervene in private property management), and of the enforceability of such policies. Centralists are more concerned with free riders in common property resource management and are more apt than decentralists to view most people as egoistic utility maximizers. Thus, the centralists see the failures of polyarchy and representative democracy as symptomatic of people's inability or unwillingness to take a rational, self-motivated approach to environmental protection. They perceive that, in a future environmental crisis, most people would want strong leadership and fast, comprehensive results, because citizens would be either apathetic or cynical about their own effectiveness at solving environmental problems. The decentralists' view of human nature assumes that, as personal involvement increases, people become much better citizens, better at problem solving and more apt to take responsibility for their actions (Warren, 1992).

This study uses participatory democracy as an evaluative criterion of environmental policy precisely because of its high standards, despite the fact that effective participation occurs more in the minds of theorists than anywhere else. As Fischer (1990) puts it:

> The virtue of democratic theory is to be found in the legacy of standards it has bequeathed. Although political theorists will continue to debate the proper application of such standards, they nonetheless provide valuable criteria against which political practices can be judged. Indeed, the very purpose of democracy is to establish a framework for engaging in open discourse and in turn, for judging its quality. As a public ideology, democratic standards are essential to the processes of honest and open discussion of public affairs. (pp. 32–33)

Using these evaluative criteria, it should be possible to draw out the essence of the potential incompatibility between sustainable environmental policies and participatory democracy. What difficulties would contemporary democracies have in achieving sustainable environmental policies, given that these policies had to be achieved for the survival of future generations? Three themes recur in the centralists' and decentralists' debate: First, both schools of thought blame existing systems of economic organization for creating environmental dilemmas.[3] Second, both groups directly or indirectly address the uses of technology and the presence—whether benign or sinister—of technocracy.[4] Third, several writers from both groups focus on the moral imperative of environmental salvation.[5] Neither centralists nor decentralists advocate diminishing generally protected human rights in favor of drastic environmental action; in fact, insofar as the discussion remains on a level of ethical or meta-ethical frameworks, it rarely prescribes a specified course of political action. But there is room, in the analytical gap left by both camps, for strongly antidemocratic alternatives.

These three themes will subsequently be called "challenges" to democracy. The task in the remainder of this chapter is to characterize these challenges in the context of sustainable environmental policy-making. How can social justice, technocracy, and economics prevent participatory democracy from achieving the kind of sustainable environmental policies described above? What might concern the democrat who wishes to maximize participation? What sorts of environmental outcomes can be expected? After exploring these potential challenges, chapter 6 assesses whether they do indeed appear to constrain participation in setting and implementing sustainable environmental policies.

The Challenge of Social Justice

This section will explore the relationship of individual equality and participation to the concept of justice. Can justice exist without democracy? Can democracy exist without justice? Is just social action, by definition, possible only through democratic participation? If just outcomes can sometimes be achieved through nondemocratic choices, are elite guardians the logical decision makers, or "choosers"?

If some people equate environmental degradation with a profound disregard for social justice and come to associate these negative outcomes with contemporary democracies, one may reasonably wonder if satisfying people's needs for justice might prompt them to insist on other forms of governance. This possibility raises several questions:

1. Does social justice require democratic politics to exist? Under what conditions?
2. Do the conclusions change when social justice is defined as environmental goods or economic goods (especially as they may conflict with environmental ones)? How so?

These questions point toward two challenges to democracy, one benign and democracy building, the other potentially antidemocratic. The more benign challenge is that social justice will be equated both with environmental justice *and* empowerment, prompting calls to reform environmental policy *toward a system that is more democratic*. This assumption is based on the following causal links:

1. Everyone has a right—even a duty—to be informed about matters affecting their environment.
2. If everyone were well-informed, an overwhelming majority of people would insist on political influence.
3. They would translate their increasing political influence—democratically—into substantially proenvironmentalist action.

Proponents of this benign challenge might concede that enough homogeneity of interests might not require a high degree of participation by everyone. There may be instances where just actions may be performed in the public interest without necessarily relying on democracy. Jane Mansbridge (1980) demonstrates that greater equality for all individuals may be superfluous if everyone shares the same goals and agenda (e.g., within a small group, or when the "trustee" model of representative democracy works best). But such situations are rare and presuppose a great deal of democratic participation, community, empathy, and trust. Thus, the benign challenge consists of a challenge to reinvigorate democratic politics.

The second challenge from social justice is one that does not necessarily equate fair environmental outcomes with democratic procedures; instead, environmental and social justice are more important as outcomes of social choice than the process (democratic or nondemocratic) that makes them possible. If democracy does not achieve sustainable environmental outcomes that can be considered fair, does reaching fair environmental outcomes require another mode of governance, one that is possibly undemocratic? The social justice challenge is strong not when interests are so concordant that few people need to bother with participation—because they know their views will be presented strongly by someone of like mind—but when our governance results in environmental outcomes we find morally wrong. We can still ask if empowerment, democracy, and social justice really can exist independently of each other; only now the focus is on whether we can imagine political outcomes that are so bad, wrong, or unfair that correcting them is more important than preserving democratic forms.[6]

Heilbroner (1980) and Ophuls (1977) both seem to suggest that survival per se will be somehow ranked in relation to political freedoms *and* that survival will prevail—barring unlikely radical changes in attitudes and behaviors—at least enough to suspend some political procedures that are considered intrinsically fair. But Heilbroner's work does not make clear whether this survival will be worth the suspension of democratic structures.[7] How much environmental justice can a nondemocratic, or very much less democratic, system ensure without political equality or a system like universal suffrage? It is possible that social justice and democracy can be separated somewhat, but they cannot be kept very far apart without the preservation of the one making a mockery of the other.

To address this question, one can search for current examples in society and politics where democratic procedures are set aside or downplayed in favor of principles or goals that ought to remain somehow outside of the political process. For example, Hochschild (1981) provides a persuasive account of why Americans are egalitarians in the political arena but much more libertarian with respect to the marketplace. She shows that most people are willing to accept differences in economic conditions on the theory that (1) differences in financial wealth are not inherently bad but may even be powerful motivators of useful labor, and that (2) economic opportunities appear to be open to everyone; what is needed to succeed is hard work and a certain amount of good luck. And even though politics and economics interact in many ways to affect an individual's influence upon the political process, the people in Hochschild's study were reluctant to bring economic and redistributive issues within the purview of particularly egalitarian democratic controls.

Lane (1986) elaborates on Hochschild's analysis, adding that

> on balance . . . it seems that the public tends to believe that the market system is a more fair agent than the political system. People tend to include the problem cases in the political domain and exclude them from the market. They ignore many of the public benefits and, with certain exceptions, prefer market goods to political goods. They prefer the market's criteria of earned deserts to the polity's criteria of equality and need, and believe that market procedures are more fair than political procedures. They are satisfied that they receive what they deserve in the market, but much less satisfied with what they receive in the polity. By a different measure, they are much more satisfied with the general income distribution among occupations than with the distribution of influence among social groups in the polity. (p. 387)

Perhaps the best example of our efforts to remove certain considerations from the democratic process is the U.S. Supreme Court, an institution insulated from the short-term ebb and flow of electoral politics and public opinion precisely so that it may be free to interpret the Constitution in the public interest. Thus, for example, the Supreme Court has developed an increasingly central role in the protection of minority rights, and sometimes, as has been the case with voting apportionment, it has found itself directly challenging participatory democracy in favor of maintaining political access for minorities (Smith, 1985, p. 128). John Hart

Ely (1980) has argued that protecting minority representation is a fundamental justification for the Supreme Court's independent judicial review of the other branches of government. Judicial review is justified because it is concerned with benefits for which there is no "substantive constitutional entitlement" (p. 145), and are thus susceptible to being hidden behind seemingly innocuous or disingenuous legislation aimed at abrogating minorities' enjoyment of these benefits. The Court has preserved these benefits by protecting not only electoral access but also opportunities for making real gains in the "pluralist bazaar" of American politics: "The whole point of the approach is to identify those groups in society to whose needs and wishes elected officials have no apparent interest in attending. If the approach makes sense, it would not make sense to assign its enforcement to anyone but the courts" (p. 151).

The best analogy to draw from the judicial example of minority protection may be that of affirmative action. With Ely, Dworkin (1985) and liberal Supreme Court justices argue that the principle of rejecting discrimination based on race transcends the democratic process. We can think of two justifications for bringing issues of minority advancement into the relatively undemocratic purview of the Court.

Affirmative Action as the Right Principle

A first argument, made by Ely and Dworkin, is in favor of affirmative action based on the fundamental principle that people shouldn't be discriminated against on the grounds of race. Because these principles are somewhat apart from the democratic process, the argument goes, it is appropriate and fair to codify the values of affirmative action into legal precedent, and regularize them into an institutional structure so that people recognize "improvement of minority representation and advancement" as one of the functions of the Court.

Of course, a court may not serve the public interest in the manner those who advocate the "principle" argument expect or desire. Such a court is presumably above the fray of interest-group politics, but in the end minorities must rely on justices to oppose the discriminatory (or anti-environmental) biases of the majority. The difficulty arises from the inescapable fact that it is this same majority that elevates these justices to the bench, though very indirectly. Now, we can hope that the "majority" is itself divided enough to exhibit a tendency to uphold minority interests—or basic values—out of self-interest, reasoning that political tides may make minorities out of majority coalitions.[8]

Pragmatically, there are serious problems in extending the analogy from affirmative action very far into environmental issues, because there is no Bill of Rights or Fourteenth Amendment to serve as a fundamental mandate (despite environmentalists' interest in seeing one adopted, as was discussed in chapter 1). So an appeal to principles beyond the democratic process is made more difficult, because we keep having to come back to a discussion of the merits of environmental "rights" or values. Currently, several states have made some efforts toward codifying environmental rights into their constitutions. Montana requires that some funds be set aside for mitigating the environmental impacts of the current generation, and Oregonians tried to put an initiative on the 1992 ballot creating an environmental bill of rights, as discussed in chapter 1. But even with these efforts, Americans will face difficulties enumerating environmental policies that have policy goals originating in the Constitution. With an environmental bill of rights, a court would have to interpret its applicability to new cases, much like courts do with the amendments covering criminal justice. As with the evolving meaning of terms like "unreasonable," the Supreme Court would still have to articulate the connection between a previously determined (but vague) right and a vehicle for some environmental benefit, say, a minimum level of bio-diversity passed down to future generations.[9]

A famous court case that illustrates the struggle to define the applicability of law to natural systems was *Sierra Club v. Morton* (92 S. Ct. 1361, 1972), popularized by Christopher Stone (1974) in an essay called "Should Trees Have Standing?" In that case, the Sierra Club sought an injunction to block a large recreational development in the Mineral King Valley of the Sequoia National Forest. The question before the court concerned the enumeration of criteria that could be used to confer legal standing. Writing for the majority, Justice Stewart specifically asked, "What must be alleged by persons who claim injury of a noneconomic nature to interests that are widely shared in order to have standing?" The Court found that "the party seeking review be himself among the injured" and suggested that the outcome of the case might have been very different if the Sierra Club had demonstrated that its own members would be personally injured by the development.

But that was not the Sierra Club's goal. Instead, the club insisted that it should act as the spokesman for the Mineral King wilderness and that the wildlife in the Sequoia National Forest had a right of its own to be heard and protected in court. The Sierra Club received a sympathetic hearing by several of the justices on the Court, notably Justice Douglas, who argued in the dissenting opinion that the Sierra Club could indeed act as guardians on behalf of nature. He added that this would not open up the courts to any and all groups who wanted to comment on actions for which they had no remote interest:

> Those who hike the Appalachian Trail into Sunfish Pond, New Jersey, and camp or sleep there, or run the Allagash in Maine, or climb the Guadalupes in West Texas, or who canoe and portage the Quetico Superior in Minnesota, certainly should have standing to defend those natural wonders before courts or agencies, though they live 3,000 miles away. Those who merely are caught up in environmental news or propaganda and flock to defend these waters or areas may be treated differently. That is why these environmental issues should be tendered by the inanimate object itself. Then there will be assurances that all of the forms of life which it represents will stand before the court—the pileated woodpecker as well as the coyote and bear, the lemmings as well as the trout in the streams. Those inarticulate members of the ecological group cannot speak. But those people who have so frequented the place as to know its values and wonders will be able to speak for the entire ecological community. (Douglas, J., dissenting, 92 S. Ct. at 751–52)

Justice Douglas's opinion does not suggest a very clear legal test for determining who should have standing as guardians. But the notion that certain groups could litigate on behalf of nature, *in a Supreme Court sympathetic to the plight of the splendid but receding American wilderness,* is a clear expression of support for the kind of guardianship presumed by the social justice challenge.

Even Justice Blackmun, who was not willing to go so far as Justice Douglas on the issue of standing, felt compelled to join the dissent. Blackmun's criterion for standing would not result in an "extensive incursion" by environmental groups into the field of environmental litigation. To Douglas's requirement that organizations "speak knowingly," he would add that an expanded interpretation of our traditional concepts of standing "need only recognize the interest of one who has a provable, sincere, dedicated, and established status." On that theory, Earth First! could have as much standing as the Sierra Club.

The most interesting point about Justice Blackmun's opinion was that he thought the Mineral King case was materially different from other issues of standing, precisely because there was something compelling about wilderness.

> This is not ordinary, run-of-the-mill litigation. The case poses—if only we choose to acknowledge and reach them—significant aspects of a wide, growing, and disturbing problem, that is, the Nation's and the world's deteriorating environment with its resulting ecological disturbances. Must our law be so rigid and our procedural concepts be so inflexible that we render ourselves helpless when the existing methods and the traditional concepts do not quite fit and do not prove to be entirely adequate for new issues? (Blackmun, J., dissenting, 92 S. Ct. at 755–56)

In essence, Justice Blackmun was asserting that environmental problems justified (1) extending the influence of the Court and (2) allowing the courts to decide who should speak for—and thus exercise potentially great power over—nature.[10] The implication of both dissenting opinions is that environmental groups would be given special powers to interpret environmental prerogatives based on, or resulting from, the dedication and knowledge of their members. Thus the courts and environmentalists could reach the "right" environmental outcomes, but the process would not be very democratic, and certainly not participatory. In order for the Douglas/Blackmun model to work, principles of environmental justice would have to be ranked, at least occasionally, over the procedural imperatives of participation.

But what is so undemocratic about the courts? Many legal scholars writing about environmental law and policy see the courts as vehicles for citizen participation, on the theory that people often cannot obtain the same, serious attention in any other policy forum.[11] But if they derived their environmental policy role from a constitutionally defined environmental "right," courts would have the ultimate authority in settling controversies arising over the articulation of that right. Conversely, if an environmental right is statutorily defined, legislatures can always overrule court decisions with new legislation (Sax, 1971, pp. 237–38).

Consequentialist Reasons for Affirmative Action

The second argument for a nondemocratic, or guardianship, model of minority protection is much more consequentialist and policy oriented. The policy reason for affirmative action, for example, might be that allowing majorities to work prejudicially against minorities leads to social unrest and is thus bad for the stability of the majority in the long run. The environmental analogy is that environmental degradation may threaten our very existence, or at least impair our standard of living to the point where the quality of our lives is unacceptably poor.

The policy argument is also prone to malfunctions, especially if there is no institutional "home," like an independent judiciary, in which the terms of the undemocratic guardian rule are deliberated and agreed upon in a fair manner. A consequentialist claim is prone to abuse by small groups or elites who insist that they "know better" and should therefore be given free reign to pursue their vision of the public interest.

The possibility that environmentalism and abuses like fascism could coexist is not so far fetched, as Bramwell (1989) demonstrates with respect to the "naturist" and ecological tendencies of Nazi Germany: "The existence of ecological ideologues among the Nazi leadership does show that National Socialism was perceived at the time as a system which had room for ecological ideas. Certainly, Nazism opposed the liberal belief, entrenched under many political labels, that nature and its laws could be transcended by human society" (p. 205). Nature was also seen as an alternative to stultifying cities that bred decay, and the Nazi "Back to the Land" program was intended to create a new peasant nobility, free of "urban intellectual homestead romanticism," in place of the old nobility:

> Apocalyptic conclusions were drawn from the American experience of the period—the dust-bowl, the apparent seizure of an advanced technological society, just as the oil-price increases of the 1970s fed apocalyptic fears of energy starvation. The specific and characteristically National Socialist argument was that second and third generation city dwellers were deemed to have lost their capability to live on the land as German peasants. Once the network of kinship and tradition had been broken it could not be restored simply by moving city-dwellers to land and telling them they were peasants again. (p. 208)

In both the affirmative action and consequentialist arguments for some kind of guardianship, there is equal support for making nondemocratic structures temporary or permanent. In either case, we can wonder how resilient democratic governance and habits may be. How slippery is the slope to a full and permanent curtailment of democratic governance? How does one distinguish between authoritarian rule and guardianship when they both may speak the language of ecological rationality? Unfortunately, this is something both the centralists and decentralists have failed to address. Dahl (1985) raises this very problem in his book on controlling nuclear weapons: "The problem now is that we have turned over to a small group of people decisions of incalculable importance . . . and it is very far from clear how, if at all, we could recapture a control that in fact we never had" (p. 7). We simply do not have a way of calibrating the breaking points of democratic trust and political habits. Ely (1980) argues that "constitutional law appropriately exists for those situations where representative government cannot be trusted, not those where we know it can" (p. 183). Could environmental protection, in light of a social justice challenge to democracy, be one of those situations?

The answer from the centralists might be a qualified yes, which is what the decentralists may have sensed and feared in the analyses of Ophuls, Hardin, and Heilbroner. The centralists found the Supreme Court model of environmental justice, based on ecological principles and long-term public interest, to be plausible; the decentralists had in mind something much closer to the guardianship model, which could subsume fairness and individual liberties in the name of increasingly personal and esoteric notions of environmental fairness. Chapter 6 will explore whether some of these visions of environmental justice played a role in the empirical studies.

The Technocratic Challenge to Democracy

From a scientific and environmentalist perspective, technocracy may seem to be a very attractive way of addressing environmental dilemmas. Environmentalism draws much of its raison d'être from science and technology because it is both informed by science and reactive to technology. In sharp contrast to the logic of liberal democracy, environmentalist political theorists of all kinds try to design or modify political structures in direct consideration of scientific findings or the effects of certain technologies (generally as reported by natural scientists). Environmentalists use science to look at the natural world in its current state and then in the state they believe it *should* exist. They take this preferred state of nature and design political governance structures so that the preferred state will be the likely outcome of political—and economic—activity. For liberal democracy, this would be putting the cart before the horse.

Indeed, the "genuine politics" Ophuls (1977) claims is so lacking consists of the power to define and create right results. R. W. Hoffert (1986) argues that this is an image of politics that makes no sense as "politics." Hoffert asserts that "the reason politics is needed at all is to deal with the historical predicament of humans in a context in which truth and authority are never finally settled because there are always multiple, competing claims regarding both truth and authority" (p. 21).

It is certainly conceivable that some interest groups might prefer technocratic policy-making to lobbying, persuasion, and compromise; the ecologists' preference may have more to do with getting the "right" technocrats in power than with philosophical inclinations for or against participation. By the same token, democratic participation can be an important rallying cry for groups that lack a strong voice.

Assuming diverse interests could agree on the principles that would guide environmental policy choices, would they know enough about the effects of various policy options to choose correctly? Society may not be able to judge policy choices to a degree of certainty that is comfortable "enough," and so environmental problems may indeed be hard for democracies to address effectively, precisely because they are fraught with scientific uncertainty and technical complexity. Such uncertainty and complexity coincides with a modern tendency to compartmentalize problems into organizations of experts and technocrats. We simply have not been able—or willing—to approach most issues within the government purview as problems of "general competency." Call it an insidious mechanism for creating fiefdoms of information, a perennial lack of civic education, or simply the nation-state's inability to coordinate hundreds of millions of citizens in democratic participation—all environmental issues recognized as problems by government are assigned a corps of technocrats.

Of course, such "technocratization" is not the special characteristic of environmental issues; most sectors of industrial societies are full of specialists and experts. But the mix of highly technical information and uncertainty creates a paradox unlike most of those found in industrial society. On the one hand, environmental issues are viewed as technical problems, and technical solutions are sought; on the other hand, there are large margins of error associated with predicting the environmental outcomes of different policy scenarios.

It is those errors that "experts" wish to eliminate through technocracy, or "the adaptation of expertise to the task of governance" (Fischer, 1990, p. 18). For example, most of the technical parameters associated with developing new technology can be known or estimated with more certainty than the parameters associated with, say, global warming models. In industrial research, experts work at marginal improvements in information; environmental problems increasingly require comprehensive understanding, because entire systems are so poorly understood.

Beyond the data gathering strengths and weaknesses raised by environmental issues, technocratic environmental management displays three attributes:

1. The ability to fill a gap of *technical competence and information*
2. The tendency for environmental legislation to require only *pro-forma public participation* of technocracies
3. The creation of *inequality and negative instead of positive authority* under technocratic management

Environmental political theory offers the above attributes as reasons for the incompatibility between environment and democracy. For example, where theorists assumed that democracy would be too "slow" (Dahl, 1990), they also concluded that organizations of experts would not need to constantly relearn enough technical information to be competent. Fischer (1990) describes the logic of technocracy at length, arguing that:

> in a highly technological society, the pivotal roles of public opinion and citizen participation are seen as artifacts of an earlier time. . . . In a governance system geared to mediating between technological and organizational imperatives and the demands of the citizenry, politics must be administratively centralized, much more technocratic, and largely elitist. Democratic government, as conventionally understood, must inevitably wither under these arrangements, a process well under way. Democracy, is, in short, taken to be an inappropriate and inferior decision-making system for the emerging postindustrial society. (p. 16)

In the centralists' work, and in much democratic theory, there is a bias against the information-processing abilities of highly participatory governance structures (Fischer, 1990, p. 22); although, given the reputations of large federal bureaucracies, it is a wonder anyone in the early 1970s thought that technocrats would bring coordination, flexibility, comprehensiveness, and efficiency to environmental management. Nonetheless, technocracy permeates the environmental policy process, so it is crucial to examine how technocracy might inhibit democratic environmental policy-making.

But if technocracy really does present a possible barrier to democratic (especially participatory) environmental policy-making, *how* does it pose its challenge? What possible elements of technocracy constrain democracy, or narrow the range of policy options considered? By what mechanisms could technocracy prevent citizens, especially nonexperts, from making and implementing decisions in environmental dilemmas? The remainder of this section examines the three attributes listed above in more detail.

Technical Competence and Information

An "informational paradox" arises from reliance on a technocracy: We create technocracies in part because we cannot imagine enough citizens becoming sufficiently active, responsible, and competent to manage the decisions we delegate to technocrats. Yet while we also expect technocrats to play a clarifying or heuristic role for society, we perceive that they make it more difficult for nonexpert citizens to understand complex environmental problems.

The emphasis on "expert" advice often requires that proposed action or topics under consideration be couched in language so technical that nonexperts may be intimidated. Similarly, experts may not take nonexpert testimony or opinions seriously. Dahl (1985) points out that a minimum level of knowledge is needed to understand the very terms on which we can safely delegate authority to experts. But there is also a tension between assumptions that governing ought to be restricted to the qualified and that an adequate level of moral competence should be widespread. For Dahl, the technical qualifications of technocrats do not overcome their most serious shortcomings for at least three reasons:

1. The specialization they necessarily acquire results in ignorance of the realm outside of the expert's field.
2. Plato's Royal Science does not exist. That is, there is no science of right—or virtuous—governance, so we cannot train or choose morally and technically competent guardians.
3. Many judgments do not rely on strictly technical assumptions.

Long before Dahl expressed these concerns, Laski (1931) stressed that expertise should not substitute for statesmanship because of the limitations Dahl cites and for these additional reasons (pp. 4–7):

1. The expert dislikes, and responds poorly to, novel views.
2. Experts often don't put the results of their inquiries into proper perspective.
3. Expertise tends to nurture elitist attitudes, or a "caste-spirit."
4. The "expert rarely understands the plain man."

Laski recognized that the "plain man" could not possibly be asked to decide policy on all technical matters but emphasized that technical ignorance was not a justification for expert guardianship: "But the inference from a knowledge that the plain man is ignorant of technical detail, and broadly speaking, uninterested in the methods by which its results are attained, is certainly not the conclusion that the expert can be left to make his own decisions" (p. 12).

While it may be attractive to "educate the citizenry by doing," increasing citizen control does not necessarily increase citizen competence. Competence may be enhanced, especially in structures like the functional representation proposals Burnheim (1985) suggests. Indeed, Pateman (1970) argued strongly that the primary function of participation is educative. When people participate in decision making in one arena (work, school), there is a crossover effect, and the habit of participation makes involvement in another arena more effective. But participation and control may not go hand in hand. We can imagine how increased control without competence could lead to a rash of despotic decisions by any group to whom control is given. The term "participation," as Pateman uses it, suggests control over actions *and* a dialogue that requires participants to offer strong reasons for carrying out a group decision.[12] If Pateman is correct about the educative function of participation, then the agencies that are poor at involving affected citizens stifle the educative process that occurs through participation. Less participation and less competence may thus go on and on in an infinite regress, further insulating technocrats from effective and enlightened public control.[13]

The timing of participation and whether it should serve to educate people or to articulate goals will also affect the role of experts. Nelkin (1977) concluded that democratization of decision making for complex technical problems requires participation at an early stage in the policy process. This is important not so much because participants can become educated early on but because social goals can be articulated to set a context for technical choices. Nelkin also finds that education is critical in improving the public's understanding of science, but more education may cause unrealistic expectations of scientists. The public often does not understand the limitations of scientific knowledge, and the more they know, the more concrete are the answers they demand. Conversely, people may reject even their own moderate involvement in the management of complex problems such as toxics when they feel that more competence is necessary (Dahl, 1990) and that technocrats really are better trained and suited for such tasks (Pierce, 1989). In such cases, the technocrat and the layperson may have a relationship similar to that of a doctor and a patient: the patient does not want to tell the doctor how to administer medical care, but in exchange for a loss of control, the patient may still require very good reasons to trust the doctor.[14]

The problems of citizen competence are further compounded by the manner in which technocracies control and disseminate information. Where an agency is the principal source of information, it has the ability to favor particular information, repress some studies, or choose not to pursue any number of research topics (Fischer, 1990, p. 111). To the extent that technocrats control how and what information is gathered and discussed, and how problems are posed, they may also define the range of policy options under serious consideration.

But unlike military departments or agencies, the bureaucracies responsible for overseeing environmental issues (e.g., forest and hazardous waste management) rarely have a monopoly on information. Today, there is almost always a credible source of technical information available with which to challenge, say, the Forest Service or the California Department of Toxic Substances Control (DTSC). Those agencies may try to control the range of information considered by delegitimizing studies and assumptions, or they may try to discount entire scientific epistemologies they suspect will lead away from their preferred goals. For example, Leiss (1972) demonstrates how the use of positive sciences predetermines policy outcomes and skews scientific results toward conclusions that will be acceptable:

> Scheler claims . . . that the positive sciences are characterized by a built-in prejudice in favor of relatively constant and uniform natural processes because those are the most useful in developing a set of techniques whereby one can predict the outcome of a plan with relative certainty and thus choose the means by which the environment may be utilized in accordance with human needs and desires. (pp. 110–11)

In contrast with the predictive agenda of the positive sciences, some of the justifications offered for changing forest management in favor of leaving old-growth groves alone or for cleaning up a site far beyond the agencies' recommendations are fraught with uncertainty. And while a risk assessment may be performed in a toxics case, agencies will be motivated to do so primarily in order to increase their *predictive capabilities and win support*. But the difference between ecologists, community activists, and technocrats may not simply be a matter of different levels of risk aversion. Community activists and environmentalists may just want to avoid any game they perceive to have such high stakes.

For example, Charles Perrow, in his book *Normal Accidents* (1984), argues that there may be a class of technologies, like nuclear power, that have low day-to-day risks associated with them but catastrophic accidental risks. These kinds of technologies "are hopeless and should be abandoned because the inevitable risks outweigh any reasonable benefits." Ultimately, Perrow stresses, the issue is not risk but the power to impose risks on the many for the benefit of the few (p. 306). Certain methods of forest management, like many hazardous waste treatment technologies (e.g., salt dome storage), are perceived to be both risky and uncertain.

Perhaps the technocrat is most appalled by a kind of rationality that is neither absolute nor instrumental, but rather social and cultural. Technical rationality relies on the mathematization of knowledge and experience, an insistence on experimental proof and the use of technically trained expert officials (Weber, cited in Fischer, 1990). The concept of "social rationality" implies that nonexperts are more comfortable living with cognitive limits (Perrow, 1984, p. 321). The socially rational person says, in effect, "We don't know what the consequences of removing all old-growth forests are, and we are uncertain even of finding out; therefore, we decline to perform the experiment." Thus, if an agency rejects "relatively inconstant and temporally unique phenomena" (Scheler, cited in Leiss, 1972, p. 110), will it also reject the groups asserting the primacy of that knowledge and information?

If the technocratic challenge to democracy is strong, the answer must be yes, especially if technocrats (inside or outside of agencies) are not pressured to alter their decisions by external actors. Of course, closing the door on the bearers of such nonpositive knowledge may be difficult if they are adept at maneuvering through the procedural blocks set in their paths by agency rules governing access and participation. Thus, groups with nonpositive knowledge must not only obtain standing, access, or a hearing, but they must also work to advance the use and legitimacy of their "different information," notably information that redirects attention to the social origins and political implications of the agency's goals.

A number of theorists, led by Jürgen Habermas, have explored ways in which technical rationality might become bounded by normative discourse (from nontechnocrats). Habermas, Fischer, Dryzek, and other critical theorists argue that face-to-face discourse aimed at broadening social understanding should emerge as the ideal form of rational political practice. For Habermas, the process of decision making is most rational not when it is engaged in choosing paths to specified ends, but when it seeks to reach communication and understanding, especially as this understanding may help to justify norms and their impacts in a social setting. The notion of "communicative rationality" espoused by critical theorists is much like the ancient Greek ideal of rationality, in which rational ideas are those that can be defended by arguments acceptable to a reasonable audience (Majone, 1988). Understanding is held to be vital because societies are organized through communication and voluntarily accepted norms (deHaven-Smith, 1988), and because norm setting defines states of the world as desirable or undesirable, thereby allowing society to agree on what constitutes policy problems. Thus, norm setting precludes, and perhaps controls, goal setting (Majone, 1988).

Critical theory posits a "public sphere" wherein an "ideal speech" situation would prevail. This public sphere is not the contemporary state itself but might include all interactions between individuals acting outside of the authority of state. In ideal speech, the search for truth is paramount;

no participants, themes, or contributions are restricted, and the "winner" to a claim of validity succeeds solely on the strength of his or her better argument (deHaven-Smith, 1988; Dryzek, 1987b).

If this theory sounds politically unworkable, its proponents have denied that it should be used to derive real-world blueprints for collective action or social manipulation (Luke and White, 1985; Dryzek, 1987b). There are several reasons for critical theorists to avoid institutional design: First, and perhaps most important, the legitimacy and success of such designs is thought to be assessed better by participants than by observers and hence could never be specified either externally or beforehand by participants themselves. Over and over, this emphasis on discursive processes is underlined. As Majone (1988) says about public deliberation, "Good arguments and open communication are not merely means to the end of efficiency, but ends in themselves" (p. 178).

Second, the circumstances making communication possible change, both to the detriment and advantage of open communication. Thus, an acceptable or legitimate *degree* of communication cannot be specified beforehand (deHaven-Smith, 1988; Luke and White, 1985). Third, and more germane to real-world applications, the amount of coercion exercised by the state cannot be specified in advance, because we cannot know how people will voluntarily recognize and accept discursive principles.[15]

But some attempts have been made to articulate the nature of discursive political institutions along the lines of critical theory. Dryzek (1987b) has listed international conflict resolutions, environmental mediation, and regulatory negotiation as a few real-world "incipient designs" that approximate discursive designs with their high emphasis on communication, process legitimation, consensus, value explication, and face-to-face interaction. Each of these designs is highly flawed if viewed against the critical theorists' ideal. International dispute resolution is often seriously hampered by the unequal power relations of participating states; for similar reasons environmental mediation often becomes cooptation (Amy, 1987), and regulatory negotiation may achieve nothing when agencies negotiate in bad faith, or when proceedings serve only to paralyze the decision making process.

Critical theorists bemoan the domination of technical forms of rationality, arguing that they preempt other ways of thinking about political goals and actions. For example, groups outside of an agency feel the need to make their knowledge seem objective, because technocracy may seem more legitimate in a climate of biased information, especially where there is a tendency to argue in favor of a separation between administration and politics. But the hope for an objective source of information is as persistent as it is disingenuous, since opponents in environmental conflicts want their perspectives legitimized but are unwilling to grant legitimacy to sources they cannot trust. To the extent that technocracies can play opponents off against each other, they may emerge looking like the wise arbiter judiciously occupying middle ground. But the referee's price may be too high: "depoliticizing" or unbiasing information may be antidemocratic if technocratic control of the policy agenda effectively closes debate on contentious policy issues. Thus, participation by nontechnocrats may become meaningless, a possibility explored in the next section.

Pro Forma Participation

If natural resource and other environmental management agencies never dealt with the public, the question of whether there is a technocratic challenge to democracy might have an obvious answer. But public agencies have been exhorted, mandated, cajoled, and threatened into establishing increasingly more procedures for public participation over the last thirty years, especially during the 1970s.

Gormley (1989) suggests that a number of causes contributed to effective institutional reforms of American bureaucracies and that these opened up agencies to public participation on an unprecedented scale. Watergate, Vietnam, new social movements, inflation, and newly perceived crises like environmental degradation provided powerful reasons for reform (Gormley, 1989, p. 32). Congress, the courts, and several administrations responded with sunshine laws (freedom of information acts), environmental impact statements, advisory committees, the Resource Planning Act (requiring public participation in federal timber sales), open meeting laws, and many other actions (Gormley, p. 45). These reforms were built upon administrative procedures agencies were already familiar with and were designed to be acceptable enough to bureaucrats that they would, at the very least, withhold active opposition. Moreover, many of these reforms were welcomed by federal and state executives and their staff, as long as citizen involvement would not undermine their goals of efficiency, expertise, and accountability (Gormley, p. 88). Gormley suggests that bureaucrats had nothing to fear: "The interest representation reforms of the 1970s allayed many of these fears by improving the quality of citizen participation in the administrative process, by reducing the need for citizens to resort to protest or litigation, and by institutionalizing representation through proxies in issue areas where this seemed appropriate" (p. 88).

Paehlke and Torgerson (1990) also argue that "the environmental issues of the 1970s have, in practice, often led to the revitalization and expansion of participatory opportunities" (p. 38). They claim that "more and more environmentalists . . . have consistently pressed at every opportunity for more and more effective means of involving the educated and the general public in decision-making processes" (p. 40). The most important legislative innovations they cite as "opening a door to the administrative state" have been:

1. Community right-to-know legislation
2. Internal responsibility systems regarding occupational health and safety
3. The use of referenda in environmental matters (p. 45)

But in light of the frustration felt by many people involved in forest or hazardous waste management disputes (see chapters 5 and 6), we can wonder if these impressive measures have gone far enough or have been effectively implemented. To the extent that participation in administrative processes has been predictable and characterized by rigid, pro forma procedures, the contemporary agency's version of participation may be a potent part of the technocratic challenge to bureaucracy. Many of the reforms listed above have become routinized, and their importance has diminished, as agencies go through the motions of public consultations, even when they have overwhelming motivation to implement a predetermined policy choice or when no participation is truly sought from the public. And more public participation does not mean increased public discretion; agencies are still bound by administrative rules and statutes, and thus cannot change certain policy decisions even if they are persuaded by external arguments. This is especially true of foresters who might wish to change the ways in which harvest sales are made so that they can offer less than the mandated timber harvest levels set by Congress.

Recalling the elements of participatory democracy discussed earlier, for participation to be effective, citizens must have some control of the agenda and some influence at the crucial decision-making stage (Dahl, 1989). Most important, the process should be educative, and viewed as credible and legitimate by all participants. As chapter 5 will show, most of the participants interviewed on the Forest Service planning process have rated it a dismal failure (Lyden, Twight, and Tuchmann, 1990), whether they were proindustry or environmentalists.

Thus many of these participatory reforms have not been unqualified successes: Environmental Impact Reports (EIRS) may not increase the rationality of administrators' decisions but only the appearance of rationality (Amy, 1990, p. 63); agencies that do not want to change their decisions may still hold multiple hearings and produce volumes of testimony without accepting substantive changes, or may adhere to the status quo because of the gridlock created by too many participants.

Paehlke and Torgerson (1990) suggest another way in which participatory procedures can backfire when imposed upon agencies. They point out that technocracy seems to require bureaucratic structures designed with specific constituencies in mind, and that it is thus very difficult to open up administrative structures to interests that have been excluded or marginalized (p. 28). This view was echoed by reports of Forest Service planners early in the Resource Planning Act (RPA) process. Planners felt inadequate to the task of conducting public meetings, sifting through public testimony, and responding appropriately (Trzyna, 1992). Whereas the industrial chemist or gyppo (logging team) crew leader both have their counterparts in the person of a chemical engineer in a toxics division or a forester with the Forest Service or the California Department of Forestry (CDF), the local activist has no "natural" counterpart in the bureaucracy.[16] That position has to be improvised or invented, often without an obvious administrative structure or "home" to support it.

Yet if the bureaucracy changes little to conform to some of its new constituents, there is also a possibility that those seeking access will model themselves more and more on agencies in order to increase their commonalities with agency officials. Paehlke (1990) asks if environmental organizations have had to adopt administrative structures similar to those they seek to challenge (as suggested by Weber), but his answer is incomplete. He points to the German Green Party as the one example of an environmentalist organization that has not adopted technocratic organizational features to pursue its program. He does not look at the many large, established environmental advocacy organizations and consulting groups, such as the Audubon and Wilderness societies, Sierra Club, and World Resources Institute, to name a few. Many opponents of environmental activists accuse environmentalists of becoming professionalized to the point of missing the inherent point of their activities, and focusing instead on fund raising and the exercise of negative authority merely to maintain political power and access. In summary, the mix of technocracy and participation may be problematic not only if technocracies resist effective involvement, but also if the act of participation becomes more important than (instead of as important as) the substantive environmental outcomes sought by participants. Paradoxically, this means that, as environmental groups narrow their scope, they improve their effectiveness at an increasingly micro policy level. This cycle continues the more they develop their access and begin to resemble the technocrats they try to influence. In their own quixotic way, the large, national environmental groups may have settled the question of participation versus environmentalism in favor of the former.

Inequality and Technocracy

Technocracy, through its affinity for technological progress, has been faulted for being dependent on inequality and domination (Borgmann, 1988; Leiss, 1972). Here, the technocratic challenge meets the economic challenge discussed later in this chapter: Borgmann (1988) claims that inequality is essential to technological progress, because technological progress—and a technocracy to manage that progress—is best served by a quantitatively expanding economy. The best "motor" for an expanding economy is inequality, especially because painful redistributive policies might have to be used to maintain an acceptable "floor" for the standard of living in a steady-state or stagnant economy. This theory rests on an assumption that technocracy is both handmaiden to and facilitator of social conflict and inequity.

William Leiss, in *The Domination of Nature* (1972), argues that the technological domination of nature is really just a form of exercising power over people: "If the idea of the *domination* of nature has any meaning at all, it is that by such means—that is, through the possession of superior technological capabilities—some men attempt to dominate and control other men. The notion of a common domination of the human race over external nature is nonsensical" (pp. 122–23). Leiss also argues that social desires and wants further drive scientific and technological progress, but with ever-increasing conflict (Borgmann's claim of inequality). As social conflict grows, technological progress and change are enlisted by all sides in their attempts to exercise dominion over others. Thus, by mirroring existing social conflict, technological change is used not to alleviate hardship and wants but to increase strife. He claims that technology does this in several different ways:

1. With the aid of technology, an economic surplus among the wealthy becomes much larger, and promises greater satisfaction of material and cultural needs. Conflict then arises over the disposition of the economic surplus.
2. Technology is fed by an ever-increasing appetite for natural resources —where these are unevenly distributed around the world, coercive measures may be sought to assure an adequate supply.
3. Imbalances among nations in controlling the external environment (especially for development purposes) act as a further abrasive influence.

Technocracy and *technology* are not one and the same, however. If the above claims about the dangers of technologies are worthy of concern, should that concern automatically be extended to conceptions of "environmental technocracy?" The answer can be affirmative if resource or environmental management agencies and professionals today rely mostly on technological perspectives and solutions to fulfill their functions, and if those technologies perpetuate inequalities.

In the case of hazardous waste management, technological progress has always played a central role, from causing hazardous by-products in the chemical, electronics, and petroleum industries to developing waste treatment alternatives (e.g., high-temperature, short-time incineration). Most of the proposals to remediate contaminated sites on the Superfund list rely on technologies of ever-increasing complexity. Indeed, many residents of low-income or ethnic communities where hazardous waste treatment technologies are often proposed would find little comfort in the knowledge that nearby sites are laboratories for state-of-the-art technical solutions. Many already perceive that treatment facilities are being sited where these groups live precisely because they have not had adequate political access. Meanwhile, state and federal agency officials far removed from these communities endeavor to explain why various treatment technologies are desirable, necessary, and safe. Thus the circle of inequality and technological progress is plausibly completed in the case of toxics.

In the case of forest management, the timber industries rely on technological progress (for harvesting and reforestation) far more than on the Forest Service, Bureau of Land Management (BLM), or CDF,[17] and innovations in information-gathering technologies increase the roles of experts and professionals who tend to be outside, and to oppose, the forest-products industry. For example, the Seattle chapter of the Wilderness Society used remote sensing and other forms of geographic information systems (GIS) to map the old-growth forests of the Pacific Northwest remaining in 1989–92. They found dramatically lower levels of old-growth acreage still standing than had been indicated by the Forest Service's own estimates, although crucial differences in the definition of "old-growth" may explain some of that variation.

It is harder to see how technology and technocrats perpetuate social inequities in the forestry case. Technological sophistication and evolution do not intrinsically fuel the demand for timber in California and the Pacific Northwest. But because technocrats keep the price of timber artificially cheap (Repetto, 1988; O'Toole, 1988) by promoting "productive," "efficient," and fast-growing reforestation based on scientific principles, building materials for housing construction in the United States remain some of the cheapest in the industrialized countries. Housing prices, however, are not so low, and thus cheap timber feeds a system of home ownership and tax advantages that is symptomatic of class inequities (between middle-class home owners and poorer renters) in American society.

Negative versus Positive Authority

It is possible that technocracy could be actively pursued by some groups in society seeking to erect barriers to environmental management they do not like. Or technocracy might be sought as a system of governance "of last resort" because of the failure of other systems. NIMBYism and the use of negative authority (the power to block action) instead of positive authority can also be options of last resort. NIMBYism and technocracy could make strange bedfellows, arising out of political gridlock and by default (in the absence of opportunities for constructive politics). This union might happen because NIMBYism creates further political gridlock. Agencies are then called upon to "do their job," being the designated resource- or pollution-management authorities able to mediate conflicts. Then, as environmental conflict leads to greater negative authority, the oppositional rhetoric calls for a diminished technocratic role in decision making, without, however, providing replacements for technocratic management. The system then comes full circle when, out of frustration at the continuation of the status quo (e.g., too much or too little timber harvesting, hazardous waste mismanagement as usual), opponents on all sides hearken back either to "the good old days" when the agencies performed their role "better," or to some idealized agency that never existed. So, when the dust of political gridlock settles, technocracy is still with us and may get another chance—albeit, with a new mantle of reforms.

But if technocracy does get "another chance," it will probably fail the tasks of ecological restoration:

> What began several hundred years ago as the technocratic exploitation of nature's resources for unlimited industrial progress today constitutes a serious ecological problem. Although substantial amounts of research and development monies are poured into the search for solutions, much of this effort appears to miss the point. By and large, it is governed by the same kind of technocratic thinking that gave rise to the problem in the first place. (Fischer, 1990, p. 44)

The Economic Challenge to Democracy

The hypothesis that modern economics constrains the democratic political process constitutes the third and final strand in the trio of challenges posed to democracy discussed in this chapter. Democratic theorists have often suggested that democracies are best sustained in wealthy societies (Lindblom, 1977). Pennock (1979) summarizes this argument by saying that higher incomes are important to democratic norms because a growing "economic pie" benefits everyone, and an industrialized, growing economy tends to have more heterogeneous interests and higher social mobility. A natural consequence is a better system of information and communication. Pennock goes on to point out that democracy requires trust (in order to share or delegate power), and that trust is a luxury that can be indulged in only by those who can afford to lose some of their material goods. Finally, Pennock notes the empirical observation that as a concern for income diminishes, people value liberty more and more (p. 218).

Dahl (1971) finds that "competitive politics" (in which he includes most democratic modern polities) requires a pluralistic social order, which, in turn, requires a decentralized economy. He is quick to point out, though, that affluence is not enough; inclusive polyarchies have their malcontents even where formally competitive politics exist.[18]

Ophuls (1977) does not dispute the possibility that democracies are most likely to exist along with wealth, but he is troubled by the strong commitment modern liberal democracies have to quantitative economic growth. He argues that our desire to avoid redistributive issues is an important reason that we keep making a larger economic pie rather than trying to cut it in different ways—ways that would be politically painful. This method of improving everyone's lot would not be so bad if it did not stir up our material consumption to such a frenzy—meaning more extraction of scarce resources, more use of energy, and more waste.

Ophuls—and many other environmentalists—would like to see a steady-state economy emerge as the solution to our present expansionist, growing economy. The steady-state economy, as Herman Daly (1980) has formulated it, would simply recycle outputs into inputs. Ideally, the steady-state economy should use renewable energy, achieve zero population growth, recycle used goods, and avoid producing nonrecyclable ones. Daly stresses that the steady-state economy could "grow" qualitatively, as people provided more and better services, and continued to research and improve on all goods without adding substantially to the material growth of society. The political issue that divides environmental political theorists is whether approximations to such a steady-state economy, or acceptable adaptations to conditions of scarcity, can be achieved with democratic governance. The literature reviewed in chapter 2 attests to the many disagreements on this question.

As mentioned earlier, the notion of physical scarcity and limits may be logically incompatible with the assumptions of liberal democracy. Liberal democrats can recognize that resources may run out and that population growth may threaten the earth's carrying capacity, but environmental political theorists and economists also speak of "social" limits to growth.

Fred Hirsch (1976) is concerned that liberalism and its reliance on the free market sets up a collision between individuals pursuing private goals and communities pursuing collective goods. Rather than focus on the frustrations and anomie of consumerism and wealth, Hirsch points out that material, productivistic growth is built in to our economy because what drives the free market is the imperative of *relative* advantage over others, or one-upmanship. He calls this the "positional economy." This kind of economics, he believes, will collapse of its own weight, so to speak, as the pie gets larger without making good on the higher rewards people are led to believe they will acquire.

Hirsch argues that "as the level of average consumption rises, an increasing portion takes on a social, as well as individual aspect" (p. 2). Furthermore, as many people achieve former peaks of consumption, the positional economy maintains itself at a small size. This causes scarce positional goods to cost increasingly more—a wasteful, inefficient expenditure of resources resulting in a transaction that has no "productive" value.

If Hirsch's analysis is accurate, the gap in economic possibilities between liberal democracy and environmentalism may be more serious than even Ophuls suggested, for it is one thing to face the politically painful process of redistribution (with which we have some experience); it is quite another problem to tackle the positional economy in order to dismantle (at the very least) the wasteful, destructive aspects of economic growth. To do so would mean redefining our notions of consumption—and how are we to tell between "good" demand and demand for absolute (e.g., basic, human), not relative needs?

Walker (1988) echoes Hirsch in arguing that Ophuls and other "neo-Hobbesians" have misread Hobbes. According to Walker, Hobbes believed the real difficulty with people as political actors lies in their thirst for "eminence," by which he means relative advantage, or power, over their fellows. Walker's argument with the centralists is that an authoritarian system would not do away with this fundamental urge, and that the creation and maintenance of scarcity through material consumption is a good way of preserving a system of eminence.

There is another aspect to the structure of modern liberal economics that has not yet been of great concern to environmentalists but has troubled democratic theorists for a fairly long time: the possibility that the very system liberal democracy depends on to generate wealth in society may in fact undermine the political ideals of democracy itself. Economic constraints on politics have been articulated differently by Benjamin Barber (1984), who mistrusts the prepolitical, nonnegotiable status of modern economic structures; by Charles Lindblom (1982) in a short essay entitled "The Market As Prison"; and by Hirsch, Robert Paul Wolff, and others.

Interestingly, these critiques point out that societies relying on decentralized "free" markets have internalized the market constraints on their political options, largely to preserve the stability of this economic system, warts and all. Writing about Hirsch's analysis, Colin Crouch (1983) observes: "The market is able to permit freedom only within a carefully limited area ringed by coercion over those who might wish to make disapproved choices as a result of their rational calculation" (p. 195). Crouch is referring to the individual's trying to make decisions based on a calculation that is rational for society as a whole but not for herself. She is punished for acting on that calculation because others will operate on individual-based calculations of self-interest. These actions Hirsch called "the tyranny of small decisions," because we are all unable to register dissatisfaction with the market in bold ways that can reverse the social impact of individual consumption. Instead, the market only allows a narrow range of options; the "tyranny of small decisions" limits our choices among them to incremental, short-term decisions (Crouch, 1983).

Lindblom's (1982) argument is similar, to the point of using the same language of "punishment." The market dispenses punishment in the form of economic downturns when it does not like the direction of government policies. Thus businesspeople need only look after their own interests to recognize disincentives for change:

> The penalty visited on them (businessmen) by business disincentives caused by proposed policies is that declining business activity is a threat to the party and the officials in power. When a decline in prosperity and employment is brought about by decisions of corporate and other business executives, it is not they but government officials who consequently are retired from their offices. That result then, is why the market might be characterized as a prison. For a broad category of political/economic affairs, it imprisons policy making, and imprisons our attempts to improve our institutions. It greatly cripples our attempts to improve the social world because it afflicts us with sluggish economic performance and unemployment simply because we begin to debate or undertake reform. (pp. 328–29)

Lindblom's point applies to environmentalism in that environmentalists may be expecting too much of liberal democracy without looking at what it is "programmed" to do. What political system would identify *as a problem* some economic activity whose reversal would threaten its own survival? To improve on liberal democracy, however, environmental political theory would have to show simultaneously that some "ecologically sound" political order could be designed that was free from such a constraint on politics, and that it could guarantee a more ecologically acceptable outcome from economic activity. Such a political order would solve the economics/environment trade-off by first recognizing that markets are not inherently incompatible with environmental protection. In effect, there need be no trade-off if social choices can be made more independently of market considerations.[19]

To evaluate the economic challenge in the two empirical cases, an important question is, Did economic imperatives restrict the range of policy options seriously discussed by advocates, policy brokers, and decision makers? For example, if people refuse to consider reduced harvesting because they fear that they will lose jobs and tax revenue and put extra welfare burdens on counties, they are concerned about protecting economic growth and viability. If they cared more about, say, libertarian beliefs about property ownership, proponents of the status quo might not use economic growth as a way to dismiss any discussion of policy alternatives. In the toxics case, facility sponsors have long argued that siting treatment facilities was indispensable to the economic well-being of oil and chemical industries (to name only two). And it was not state and federal decision makers who began NIMBYism; several sponsors across the country did in fact obtain permits to site facilities only to encounter public—not official—opposition. Before NIMBYism became widespread, it is unlikely that an agency official or county supervisor would have blocked a permit for a landfill sponsored by powerful local and regional economic interests.

It was probably economic considerations that motivated agencies like the Forest Service and the Bureau of Land Management to manage more forests for timber than other uses. Indeed, in a well-publicized federal case forcing the agencies to protect spotted owl habitat, a district court judge enumerated 14 "economic and social consequences" to reducing harvest levels on behalf of the spotted owl.[20] Deciding in favor of an injunction barring logging until a plan to protect spotted owl was in place, Judge Dwyer found:

> Any reduction in federal timber sales will have adverse effects on some timber industry firms and their employees, and a suspension of owl habitat sales in the national forests is no exception. But while the loss of old growth is permanent, the economic effects of an injunction are temporary and can be minimized in many ways. . . .
>
> To bypass the environmental laws, either briefly or permanently, would not fend off the changes transforming the timber industry. The argument that the mightiest economy on earth cannot afford to preserve old growth forests for a short time, while it reaches an overdue decision on how to manage them, is not convincing today. It would be even less so a year or a century from now. (p. 34)

Judge Dwyer did not, however, assert that any unusual precedent was set by his decision; if he refuted economic arguments against issuing an injunction, it was because the National Forest Management Act of 1976 (NFMA) required that noneconomic considerations be taken into account in forest management, and because the Forest Service had improperly considered economics when it disregarded the law.

Perhaps the most telling finding in the case went to the heart of why the Forest Service was not fulfilling its mandate:

> More is involved here than a simple failure by an agency to comply with its governing statute. The most recent violation of NFMA exemplifies a deliberate and systematic refusal by the Forest Service and the Fish and Wildlife Service to comply with laws protecting wildlife. This is not the doing of scientists, foresters, rangers, and others at the working levels of these agencies. It reflects decisions made by higher authorities in the executive branch of government. (p. 20)

The Dwyer decision brought back policy options that had been swallowed up by economic considerations and constraints. But if Forest Service technocrats were not to blame for the service's mismanagement, Judge Dwyer's opinion suggests that politics-as-usual at the secretary and cabinet levels governed decision makers in the agencies. And whether they were breaking the law or not, the political appointees in the agencies are entitled to view their party's electoral success as a mandate to carry out their resource management philosophies. These may, in turn, be more attuned to the economic benefits of publicly owned natural resources than are other management philosophies that have fewer champions in the executive branch.

And there is reason to fear that technocrats will be especially interested in advancing economic prerogatives in their resource management ideologies. In such cases, the economic and technocratic challenges combine, as Fischer (1990) argues:

> Planners and managers blur the distinction between the worlds of economic production and social interaction, thus making it difficult for many to distinguish between the priorities of the economic system and those of their own lives. It is not that people should reject economic and technological progress, but rather that they should establish their own relationships to it through the processes of intersubjective discourse. (p. 47)

CHAPTER FOUR Empirical Study of the Centralist-Decentralist Debate

Large-Scale Survey Methods

We cannot conduct an experiment in environmental crisis, and so we must turn to the policy preferences of those involved in environmental disputes to shed light on the roles that the three challenges to democracy—social justice, technocracy, and economics—may play in environmental policy participation. But have we learned much about the implications of the centralist-decentralist debate from empirical work, particularly in the environmental survey literature? What has it told us about trade-offs between economic cost and environmental protection, or between administration and participation?

Many of the questions asked of this study's respondents were designed to probe the nature of democratic principles among people directly involved in environmental controversies. Did respondents downplay the importance of participation when faced with serious environmental threats? Or, conversely, were environmental values deemed inferior to political participation or a strong economy? Did respondents view participation and proenvironment policies as somehow incompatible? Did the term "participation" evoke positive or negative reactions precisely because opponents equate "participation" with certain outcomes, favorable or unfavorable to their cause? That is, can they only imagine participation as a conflictual process, or can they envision the possibility of fair procedures arrived at by consensus?

These questions are too complex to answer through large, quantitative survey methods, but approximations can be obtained by reviewing the attitudes and behaviors of people involved in environmental controversies. For example, Pierce, Lovrich, and Matsuoka (1990) surveyed Japanese and American publics, activists, and policy elites to determine what governed the support of these respective groups for citizen participation in environmental politics. Pierce et al. wanted to know the effect of four kinds of individual variation on support for participation. These were background attributes, policy-relevant knowledge, political orientation, and policy-specific issues of special local importance to respondents.[1]

Without controlling for these sources of individual variation, Pierce et al. found that most policy elites and activists supported citizen participation in environmental politics. When measured on a scale of 1 (no value) to 7 (great value), 54 percent of the activists, 54 percent of the public, and 78 percent of the policy elites in the United States chose a value between 5 and 7, indicating strong value for participation. Pierce et al. also found that the personal attributes of education, gender, income, and age had little influence on support for participation; indeed, it remained quite high in all three groups. The policy-relevant knowledge variables showed consistently high support for participation, although more so among activists and elites than the general public. Generally, more familiarity with technical terms and issues, local environmental concerns, and understanding of ecology correlated with increased support for participation.

The variables of general political orientation revealed few surprises also: higher support for the new ecological paradigm (NEP),[2] adoption of postmaterialist values, and preservationist affinities all led to increased support for citizen participation—although less so for the general public than for activists and policy elites. This might imply that the general public has a slightly more "centralist" attitude about environmental management than do activists and policy elites, possibly because policy elites and activists are more accustomed to the effects of their own participation and anticipate that their "policy competence" rises as they acquire more responsibility.

Perhaps the most interesting and relevant observation, for the purposes of this study, concerned locale-specific beliefs about environmental policy. Pierce et al. found that "the greater the danger perceived, and the more serious the perceived problem, the greater the support for public involvement across publics, activists and elites in both the United States and Japan," (p. 52). This result flies in the face of Heilbroner's (1980) and Ophuls's (1977) implications. They both argue that as people perceive worsening environmental conditions, they seek "stronger" leadership in the form of centralized or authoritarian control. A true test of Ophuls's and Heilbroner's predictions may not be possible, however, unless large populations perceive themselves to be threatened by global environmental crises.

While the Pierce et al. study is encouraging for democratic environmentalists, we are still unsure of its implications. The authors write that "the expectation is that greater knowledge will be associated with support for citizen involvement." It is important to note that they mean that they expect individuals with greater policy-relevant knowledge to seek more involvement *for themselves*.[3] We don't know from their study whether respondents' support for participation was comprehensive, even to the point of greater inclusion for opponents. We are especially in the dark when pondering what people mean by "involvement." Do they mean advisory committees, more turnout at elections, or something else altogether? Moreover, there is no way of telling if these respondents linked their policy-relevant knowledge to outcomes: would greater policy-relevant knowledge lead not only to support for participation but also to support for proenvironmental policies? And would proenvironmental policies, in turn, lead to substantive environmental changes "on the ground?"

Some of these questions are addressed in another study by Pierce (1989) that further examines the level of policy-relevant knowledge on environmental issues held by people in Shizuoka, Japan and Spokane, Washington. Interestingly, Pierce suggests "that the major portion of the linkage moves *from* participation *to* knowledge" (p. 84). In his view, "individuals perceiving citizen participation to be of greater value are more knowledgeable" (p. 81). Did participation increase these respondents' knowledge, or were they informed and active participants throughout their public and private lives? Pierce's study does not give a rich account of the effect of participation on his respondents but certainly does support decentralist notions about transformation and democracy. But Pierce also wanted to know if knowledge acted as a barrier to participation and found that it did not, for two reasons.

First, knowledge was not limited by background and could be acquired independent of one's standing in the social structure (p. 92). The respondents who had high levels of information were not all policy elites, highly educated individuals, or members of high-income classes. This finding tends to discount the notion that only experts, technocrats, and elites can acquire the kind of information necessary to participate effectively. Pierce's study, as well as an analysis by Fortmann, Everett, and Wollenberg (1986), of timber harvest plan protests confirms that proximity to an environmental threat or degradation is a better predictor of concern, policy-relevant knowledge, and activism than class or urban versus rural residence. Thus, diverse residents around Suruga Bay in Japan were active and informed because they feared mercury poisoning; and rural residents in northern California were more likely to protest a timber harvest plan that might degrade local water quality than outside environmentalist, urban agitators "who protest everything" (Fortmann et al., 1986, pp. 6–7).

Second, if citizens had sufficient motivation to participate, they could acquire policy-relevant knowledge and begin to think about common environmental interests rather than merely self-interest (e.g., limiting industrial development around Japan's Suruga Bay or protecting groundwater in Spokane).

These findings would suggest support for participatory mechanisms of environmental policy-making, but one such effort—citizen involvement under the Resource Planning Act (RPA)—closed with very mixed reviews. The RPA mandated a system of public consultation on Forest Service planning processes. Lyden, Twight, and Tuchmann (1990) offer a glimpse of the Forest Service's performance by analyzing survey responses from Service personnel and public participants.

Two findings are the most striking. First, the authors found that a majority of the public respondents (60.4 percent) wanted to comment on *all* aspects of the planning process. Fewer in the Service—50.5 percent—found it important that the public comment on all aspects of the process. Interestingly, most of the public respondents were not inclined to comment on whole timber plan drafts: 45.1 percent would prefer to comment on simplified plans, 24 percent said that "special simplified statements should be aimed at the interest of each particular client group," and only 27 percent felt it necessary to comment on the whole plans. For lack of a follow-up, we cannot tell what "all of the process" meant to each respondent.

A more serious finding, however, was that more than 50 percent of the public respondents who had participated in RPA planning said that the Forest Service had made *no* changes in its plans in response to public participation. Eighteen percent of the public said that some changes had resulted because of their participation, and *less than 2 percent* could respond that a lot of change had occurred. Despite this weak show of support for the RPA process, 77 percent of the public respondents said that they would participate if the process was effective in turning their involvement into policy changes (Lyden et al., 1990, p. 136).

If large-scale surveys seem to indicate that participation is generically desirable (who could be against "more democracy?"), they similarly indicate strong support for environmental protection "despite high costs" (Dunlap, 1991a, 1991b).[4] But these surveys are poor instruments for probing opinions on the balance between economic growth and environmental objectives. Some survey research on environmental attitudes and behavior has taken up the possible conflict between environmental protection and economic growth, but few studies have tested how preferences for increased costs or democratic governance *in environmental issues* hold up under different economic scenarios. Indeed, the paradox in many surveys that ask questions about economics and environment is that a majority of respondents agrees that more money (sometimes "as much as is needed") should be spent protecting the environment, but they consistently rate other programs above environmental protection as policy priorities.[5] And although the degree or type of sacrifice is usually unspecified in surveys, this trend suggests that respondents faced with actual trade-offs may wish to address policy issues of more immediate concern. Or it may be that the economy/environment trade-off is really very abstract in people's minds when they are asked to judge if the two are incompatible. Perhaps respondents perceive environmental costs to involve reshuffling of governmental budget priorities—in essence, listing money already appropriated for something—rather than making painful choices that would affect personal incomes.

Trees and Toxics[6]

In research more directly related to the two issue areas in this study, polls have shown strong support for jobs and economic growth when pitted against forest preservation or the survival of the northern spotted owl. A typical example of localized responses is a poll commissioned by the *Oregonian* in 1989 and 1990. That poll asked respondents in Oregon: Which, if any, is the most important use of the state's forest lands in the state as a whole? In both years, well over half of the respondents strongly favored economic and industrial values.[7] Approximately half of the respondents agreed that "cutting trees causes serious damage to fish and wildlife." However, the large majority (two-thirds) did not think cutting trees was a major source of water pollution and were convinced that logging kept forests "healthy and productive" (74–76 percent). An overwhelming majority (87–88 percent) believed that logging was a major source of jobs and revenue.[8] The poll did not explain whether respondents thought that timber was "a major source of jobs and revenue" in local communities or the state as a whole. There is thus no way to tell if respondents were thinking locally or were misinformed about their state's economy.

On a related topic, the *Oregonian* asked whether "to protect the spotted owl, we [should] stop logging on large tracts of federal timberland as recommended by the recent federal study, even if it means a loss of jobs?" Sixty percent of the respondents said no. Thus, Oregonians, to a large extent, are strongly concerned about jobs and economic prosperity when confronted with questions of forest management, even though the timber industry directly employs less than 10 percent of the Oregon workforce.

If such a small part of the workforce is employed in the forest products industry, who are the champions and opponents of logging? A nagging deficiency of much survey and polling data is that respondents are often not identified clearly enough for us to know which groups and individuals hold opposing positions. The study by Fortmann, Everett, and Wollenberg (1986) goes a long way toward rectifying this data gap in research on timber harvest plan (THP) protests in California. In their research, Fortmann et al. analyzed letters written during 1977 to 1985 to the California Department of Forestry (CDF) concerning approximately 2,000 protested and unprotested THPs in four regions of the CDF. They were interested in finding out how much protest there was, whether protests were on the rise, why THPs were protested, who protested them, and where protesters lived. Perhaps their most interesting claim was that "neighbors" (people living near the THP site who were not activists) accounted for more than two-thirds of all written protests filed with the CDF. Fortmann et al. asked local CDF officials which individuals had reputations as activist environmentalists and scanned letters for their names. However, they did not define "community groups" in their article, so it is difficult to tell how accurately they depicted the identity of protesters.

Fortmann et al. also found that protests were more likely if THP sites were located in standard metropolitan statistical areas (a classification based on population density) and if there was no national forest in the protesters' county. The most frequently protested issues were those directly affecting property rights: "Neighbors of timber harvesting often want to be assured that their roads will not be damaged, that their water supply will not be polluted or destroyed, that their children will not be hit by logging trucks, that trees on their property will not be cut, and that the area around their property will not be degraded" (p. 28). Finally, not very many THPs were protested; only 2 percent of all plans were protested in the period between 1977 and 1985. However, protests were clustered around 1985 rather than evenly spread out over the period studied, suggesting that THP protests in the mid-eighties were on the rise. If Fortmann and her colleagues are correct, "outside agitators" are not primarily responsible for criticizing natural resource management in California's private forests, but there may be a stronger urban bias against destructive harvesting practices than would be found in traditionally rural areas.

One of the best recent studies of participation in hazardous waste management policy was conducted by Mazmanian, Stanley-Jones, and Green (1988). Using both in-depth interviews and questionnaires, they surveyed 50 participants in the Southern California Hazardous Waste Management Authority, the Contra Costa (Hazardous Waste) Task Force, and the Hazardous Waste Management Council. A number of their findings mirror those of this study, notably on the following issues:

1. The threats to public health posed by toxics
2. The value of local control
3. Whose views should be included in toxics debates
4. The major goals of participants
5. Participants' definitions of "success"

All of these topics were also taken up during interviews conducted for this study and thus serve as a good check on the findings described in chapter 5. Not surprisingly, respondents thought that threats to the drinking water supply were "very serious" (63.3 percent), as were threats to air quality (34.7 percent), and the industrial work environment (44.9 percent). In short, hazardous waste was an issue principally because of its threat to human health. Respondents in the toxics case in this study echoed that concern for human health above any others (e.g., recreation, agriculture, economic growth).

Mazmanian also found that respondents placed a great deal of importance on participation of diverse groups in hazardous waste management and on local control. Indeed 73 percent agreed that the best government is local, for two reasons: first, because local control could be useful in fighting off state or federal preemption and, second, because local control—and by implication, public involvement—is what people in a democracy should pursue. As one of their respondents put it, "It's sort of returning to some of the fundamental notions of self-government that the free society affords us the opportunity to do" (p. 71). Mazmanian and his colleagues did not ask respondents if local control produced more desirable environmental outcomes. That question was central to this study, because it indicated whether respondents favored local control for reasons of self-fulfillment, environmental benefit, or both. The study did find, however, that respondents rated the participation of business, environmentalists, health organizations, and officials quite a bit higher than that of "ordinary citizens." It was unclear, however, whether "citizens" who participated would be ranked in with "environmentalists" or "business," and thus it's difficult to tell whether respondents thought the relationship between NIMBYism and participation was constructive.[9]

Given that health risks were deemed the most important threats posed by toxics, one would expect these respondents to define their goals in terms of health protection. Mazmanian et al. found that the highest number of participants (60 percent) ranked "eliminating land disposal of hazardous waste" as a "goal of many participants," along with "locating new sites for treatment and storage" (54 percent) and "reducing hazardous waste generation" (54 percent)—far above "eliminating health risks" (36 percent). Different goals also led respondents to view collaboration differently. Those who sought lifestyle changes favored a collaborative, deliberative policy process, but those who were most eager to site facilities were frustrated by the Waste Management Authority's consultative process.

As in the present study, Mazmanian et al. asked the participants of the three groups questions about what the respondents called success. They found that "building understanding" and "flexibility" were as highly rated definitions of success as "cleaning up Superfund sites." Perhaps their most important finding about "success," for the purposes of this study, was that the final outcome of the collaborative process really did influence respondents' views about success. This implies that, in the minds of some participants, a fair process could not exist independently of its desired outcome:

> In other words, what happened in the process clearly made a difference to the participants' ultimate assessment of its effectiveness as an appropriate forum for decision-making. One can infer that if the participants had a bad experience, they would be equally harsh in judging its effectiveness as a way of addressing a public-policy problem. (p. 80)

What We Do Not Know from Opinion Research

The studies just described shed light on just a few aspects of the topics described in the previous chapters. Important questions about the prospects for participatory democracy and democratic conflict resolution, the role of technical information and citizen participation, and the place of economic growth in people's political outlooks, however, still remain largely unanswered.

Unfortunately, much of the national soul-searching over environmental attitudes has been conducted at a typically superficial level, as if to reflect the sound-bite treatment of news so prevalent since the early eighties. Most survey research does little to describe *who*, in fact, is thinking or behaving like an environmentalist and who isn't. Cotgrove (1982) attempts to separate respondents into identifiable groups by characterizing opponents on environmental issues as primarily "industrialists" or "environmentalists." He then uses this shorthand to probe some of the socioeconomic and political values that define their antagonism in the political arena. But as useful as Cotgrove's surveys are, they only hint at the questions raised in this study, because he does not probe the *degree* to which political participation is valued, especially in light of possible trade-offs with environmental benefits.

It is certainly useful to learn that identical percentages (42 percent each) of the public, environmentalists, and industrialists assigned a "high priority" to the policy value "giving people more say in important government decisions" (Cotgrove, 1982, p. 48). Yet this finding is fairly limited in its implications, especially when taken out of a controversial context. It is easy to support "having more say" as an abstract notion, but once again, it is unclear what the respondents surveyed meant: Did they have some conception in mind of what it meant to participate, or are the terms "involvement, participation, and greater say" generic even in the minds of respondents? What price would respondents actually pay in order to achieve a "greater say"? There are costs in information, time, and more important, in the power of entrenched interests—how would they make these trade-offs?

The findings of opinion researchers do generally support the conclusion, made in chapter 5, that technical complexity and information are not seen as legitimate barriers to participation. But what of the transformative effect of participation? Laypersons who become active and informed join advocacy coalitions that are, in turn, composed of policy elites—even if they are not from upper socioeconomic strata of society. The transformation of disinterested layperson to activist or policy elite involves extensive policy learning. Pierce's research (1989) does not account for how long that learning takes—whether acquiring policy-relevant information, or disseminating it through a community, takes a relatively long or short time. For most of the respondents in this study, a minimum of one to two years were necessary to become sufficiently informed and interested to take action. How does this "time and information cost" interact with participation and policy? Are there some forms of participation that make long, entrenched policy disputes worth the effort, and others that do not?[10]

Perhaps the most disappointing thing about the studies described here is that they do not link attitudes to environmental policy outcomes. Very little in Dunlap's summary suggests that pollsters have had an interest in determining either the extent of proenvironment behavior, or whether (and how) proenvironmental public opinion had any implications for political institutions.

The empirical findings of chapter 5 bridge the gap between theoretical claims about what actors and institutions can and should do about environmental policy, and what people actually know about environmental problems, or their preferences about different states of the world. The rest of this chapter describes the in-depth interview methodology used in the old-growth and toxics controversies.

Empirical Methodology

As noted, it is very difficult to infer the links between variables such as participation and environmental outcomes in survey research. Survey research also suffers from being conducted without special attention to the *context* of the respondents' experiences, yet it may actually taint responses by prompting opinions, controversies, and topics that would not otherwise have been naturally forthcoming (Babbie, 1986, pp. 232–33). By and large, survey research better serves the value of generality than validity, because large sample sizes permit statistical verification at the cost of in-depth and follow-up questioning (and a good deal of just plain listening). Generality was one of the reasons that some survey research was summarized in this chapter. If the direction of the findings in this study follows that of published survey research, we might be more confident of their generality, to the extent that there was substantive overlap between the topics of the different modes of inquiry.

In contrast to survey research, intensive qualitative interview methods allow the respondent to make the links between different variables (and, perhaps unknowingly, to suggest variables new to the researcher) as he or she sees them (Hochschild, 1981, p. 25). Moreover, in-depth interviews enhance the possibility for clarification and elaboration on points respondents make. In-depth field research can also help identify promising new approaches and topics for further study. Thus the research for the two case studies herein relies on in-depth interviews of respondents rather than on survey questionnaires administered to large numbers of respondents.

It is also important, however, to know the limits of the in-depth interview method: it cannot be used to predict the behavior of the study's respondent groups, partly because of the small pool of respondents and partly because no statistical analysis can be performed that can establish firm trends in larger populations. In-depth field research may also suffer from a lack of reliability if the researcher's descriptions exhibit undue bias. In such cases, subsequent studies may show very different results even if the same methods are used.

Nevertheless, intensive interviews are ideal for this study because they can generate insights into the preferences of respondents, given certain policy alternatives which, to the extent possible, the respondents generate without much prompting from the investigator. While there may be a loss in generality, the study's applicability can be based on its resonance with other studies, quantitative survey trends, and research experience in similar cases.[11]

A final word about the choice of in-depth interviews. Is it inappropriate to probe individual attitudes and preferences to make inferences about the accuracy of what might be termed "structural arguments"? That is, if the centralist-decentralist debate concerns itself with the logical implications of choices between institutional structures, are in-depth interviews of participants the right sort of tool for measuring such claims?

While it is impossible to get people to talk about system- or structural-level implications, we can determine if the preponderance of evidence from the environmental cases used here suggests trends in the evolution of institutional structures and associated political paradigms. Choosing self-described participants ensures that responses will have high salience. People who are not involved in these issues are not expected to have much to say about the values of, and reasons for, participation.

Centralist and Decentralist Hypotheses

Chapters 5 and 6 will explore several of the more important propositions made by the centralists and decentralists that are described in chapters 2 and 3, especially the claims made about the role of participation in environmental policy. At least six hypotheses based on the centralist-decentralist literature can be posited as empirically testable propositions:

1. If people perceive that they are in an environmental crisis, they will seek centralized leadership.
2. People perceive that local control over environmental management results in more desirable environmental outcomes than central control (e.g., state and federal).
3. People who participate in the policy process over environmental dilemmas judge the value of their experience by what happens in the physical environment.
4. People who participate intensively on environmental issues gain more control over environmental policy outcomes.
5. Technical complexity is used by experts to block or negate public participation on environmental issues. Similarly, nonexperts affected by environmental controversies limit their own participation, because technical complexity intimidates them and prevents them from identifying and effectively advancing their goals.
6. Concerns over economic growth sharply constrain the scope and number of distinct environmental policy options that are seriously considered by participants in the environmental policy process.

These hypotheses embody the centralist and decentralist concerns about the right balance of participation and environmental outcomes. By exploring these hypotheses, one can infer whether people involved in environmental controversies see their options in the same ways as do the environmental political theorists discussed in chapter 2.

Each hypothesis relates to one or more of the challenges to participatory environmental policy-making described in chapter 3. Hypotheses 1 through 3 relate to the social justice challenge by making claims about the effects of participation on both environmental systems and political fairness, and about the legitimacy of the decision-making process. If hypotheses 1 and 3 are false (or incomplete), then some people may concede that socially just environmental policies might be decided best through nondemocratic or guardianship processes. Hypothesis 2 has bearing on both the social justice challenge (if people perceive it to be true, they are likely to favor increased participation for themselves) and the technocratic challenge (if true, people do not perceive experts remote from their everyday experience to be good environmental policy decision makers).

Hypotheses 4 and 5 relate mostly to the technocratic challenge; if 5 is perceived to be true, then the technocratic challenge is strong indeed. Similarly, if 4 is false, then one has to explore reasons that participation is not effective. Is it because technocrats block people's gains in policy control, or simply because, as is often the case, the most direct interaction occurs with policymakers and political structures that have little influence over the issue in question (e.g., county supervisors pressured by constituents to effectuate policy change in nearby national forests)?

Hypothesis 3, along with 6, also relates to the economic challenge, particularly if participants judge the value of their involvement by how it affects economic growth and jobs. If hypothesis 6 is true, the economic challenge can be considered a constraint on democratically determined policy options.

The Empirical Cases: Forestry and Hazardous Waste Management

We will use two very different kinds of environmental problems—hazardous waste management and old-growth forest harvesting, both represented by a range of environmentalist philosophies—to assess the degree to which different issues pose dilemmas for policymakers and political institutions. The categorization of environmental problems has political significance because not all of them are perceived to be equally threatening. As is true with many problems, the greatest resources will be devoted to the ones that present immediate threats.

Hazardous waste management (toxics) in California was chosen as the first example because it represents the claim that "the only important environmental issues are human health issues." The extensive public involvement in toxics issues also provides a laboratory to test these hypotheses. Hazardous waste management is also a bellwether issue: trends in participation over toxics issues may presage patterns in other environmental controversies. Toxic pollution has received a great deal of attention at all levels of the California economy and government largely because of highly publicized toxics crises and vigorous opposition mounted by citizens against landfills and treatment facilities near where they live and work (commonly known as the NIMBY—Not-In-My-Backyard—syndrome).

The second example is the battle to save old-growth forests throughout the Pacific Northwest, focusing on northern California and Oregon. In the case of forests, human health is, at most, a vague concern caused by projections of ecosystem degradation years in the future. Forests also grow back when replanted, although not to their previous complexity, and, arguably, not to their previous ecological value. Thus, it has been harder to call logging a "problem" for both economic and health reasons. The social and political polarization in the old-growth debate is very different from that of toxics, which leads to the question: Can the politics of two very distinct environmental arenas shed light on the basic issues of democracy and environmentalism under investigation in this study? If they can, as is the premise of this study, do the toxics and the old-growth conflicts affect democratic governance differently?

To answer these questions, I conducted over 50 interviews, in person and by telephone, over an 11-month period. Interviews generally lasted from one to two hours. The respondents were chosen because they were involved in one of the two environmental cases. "Involvement" included living in a neighborhood affected by hazardous wastes (either a site cleanup or treatment facility), holding a full-time job affected by local environmental controversies, or having a position directly related to environmental issues (e.g., in a resource agency or as a lobbyist).

Twenty-eight interviews were conducted for the forestry case and 27 for the toxics case. The largest numbers of interviews were conducted with environmentalists, officials and personnel of various industries, and agency staff (mostly state and federal). Legislators were difficult to gain access to; consequently only three interviews were obtained.

Respondents were chosen largely by a snowball or networking method

Table 4.1 Number and Occupation of Interview Respondents

	FORESTRY	HAZARDOUS WASTE
Environmental Activists	7	6
Forest Products Industry	8	
Oil and Chemical Industry		4
Waste Management Industry		3
Legislator	2	1
Legislative Staff	3	5
Natural Resources Agency	5	
Toxics Agency		7
Labor-Environmental Coalition	3	1

whereby initial interviews led to expanded lists of names and contacts. Respondents were often asked to name someone who they knew could articulate an opposing point of view. On several occasions, a respondent would give several names of people in their "policy network." Some respondents were identified by articles they had written or as sources in news reports. Table 4.1 summarizes the number of interviews conducted and the categories of respondents used in both cases.

In the forestry case, interviews were conducted with loggers, small and large owners or representatives of the forest products industry, Forest Service officials, labor representatives, an industry trade official, a California state assemblyman who has written many timber reform bills, and local or state-level environmental activists. Six of these interviews were conducted by Jerry Moles for the Klamath Bioregional Management Project. While attending several meetings of the Klamath Project, county supervisors from northern California were interviewed, as well as officials from the federal Bureau of Land Management, the state departments of Fish and Game, and those of Forestry and Fire Protection.

In the "toxics" case, interviews were obtained with representatives from a large chemical company and several hazardous waste management firms; chemical and waste management industry lobbyists; neighborhood environmental activists and organizers in Sunnyvale, Livermore, and Kettleman City, California; a county supervisor who has been instrumental in regional hazardous waste management commissions; staff specialists on toxics in the state legislature and Congress; and officials of the federal EPA and the California state Department of Toxic Substances Control (DTSC), now part of the new Cal-EPA.

Links to Theoretical Concerns: The Challenges to Democracy

The hypotheses described above were a challenge to measure directly. I chose questions that would draw out the respondents' views on political participation, environmental outcomes, and the role of technical information. From their answers I gauged whether they believed political participation and "desirable" environmental outcomes were competing principles or if their importance changed with different circumstances. It was especially important to know if respondents attached intrinsic value to open participation, environmental protection, or both. If either of these was considered expendable or negotiable—in favor of the other principle—the answer could explain much about how robust participation and sustainable environmental protection really were for each participant.

To elicit views on local versus centralized environmental management, I asked respondents if they would prefer to see local communities have more of a say in national forest management or waste management. A follow-up question often asked if local control would have an effect on the environmental goals people valued different from that of a more centralized approach, as with a federal or state agency.

Asking a direct question about the relative importance of participation and sustainable environmental protection risked leading respondents into ranking the two. To avoid this, some questions asked people what they called "success" and what they would like to see happen over the next 10 or 20 years. Most respondents needed no prompting and had very specific goals in mind. Their answers indicated whether they placed a premium on procedural outcomes (political access) or physical outcomes (site cleanup, more logging). Some responses were mixed, where, for example, procedural success was considered to be contingent on a minimum level of success "on the ground."

Did some people favor democratic participation over bureaucratic control for pragmatic reasons? Or was democracy perceived to be a slower process of implementing environmental policies than other means available? To shed light on these questions respondents were asked if, given

Table 4.2 Theoretical Hypotheses and Sample Questions

HYPOTHESES	SAMPLE QUESTIONS
1. Crisis Leadership	Should locals have more of a say in forest or toxics management? Why have you become so involved?
2. Local Control = Better Outcomes	Would locals take the best care of the forest or hazardous waste?
3. Outcome Determines Legitimacy	What do you call "success"?
4. Participation Yields Control	How effective have you been through your involvement in this issue?
5. Technical Complexity Limits Participation	How much information do you or other people need to have to participate effectively? Where do you get your information?
6. Economics Limit Deliberation	How should we pay for your preferred policy, goals, programs?

all the ways people have of protesting a decision or of getting involved in making one, they thought that *more* public involvement made reaching decisions (that they found appropriate or correct) more, less, or equally time-consuming.

Centralist and decentralist theorists alike have long been concerned that technical complexity effectively makes environmental problems inscrutable to the average layperson—so inscrutable, in fact, that citizens are both discouraged about and intimidated from even trying to influence environmental public policy in a way that might lead to more sustainable environmental policies. To illuminate these theoretical concerns, respondents were asked, "How much technical knowledge is needed in order to participate effectively?" and, "Whose views do you trust, and who has credible information?"

Opinions on economic priorities were generally elicited by asking respondents what it would cost to reach their goals and who should pay. Often a simple question of How do we get there from here? was enough to generate discussion of economic constraints and alternatives, because economic issues were everpresent in respondents' minds as they contemplated difficult implementation tasks. Table 4.2 summarizes links between centralist and decentralist hypotheses and questions used in respondent interviews.

It should be noted that respondents often raised relevant topics without being prompted by questions. Thus, although questions were structured so as to bias answers as little as possible, respondents' views were often gleaned from their uninterrupted and free-ranging discussion.

CHAPTER FIVE Trees and Toxics

The Age of Ecology is surely transforming American politics in a number of key areas. Responses from these interviews show that many first-time activists now have to face the limitations of public participation. Individuals who often participate in environmental controversies don't report making significant gains in control or influence over policy. Despite two decades of "sunshine laws" and public hearings, rank-and-file citizens seem to be losing their faith in the democratic wish—a dream of a land of enlightened citizens and meaningful participation steeped in the rhetoric of participatory democracy: community, town meetings, accountability, and self-determination.[1]

Perhaps as a result of such disillusionment, environmental politics are becoming more and more pragmatic. Thus while respondents to this inquiry differed in the value they placed on local versus central control, on the whole, environmentalists sought community control of environmental policy when they succeeded in showing how their own neighborhood or town suffered from environmental degradation. In this way, when public health and social justice were equated with environmental protection, environmentalists sought political decentralization. Some reformers want to use environmental regionalism to usher in such decentralization; others argue that changing formal jurisdictions is impossible without altering the balance of power in natural resource conflicts.[2]

But what is the end point of all this participation? What are people learning about their powers, what political habits are they acquiring? If social justice and local power are becoming increasingly important, should we expect respondents to rank these as ultimate goals? Evidently not: participation may have its limits, but for most respondents, the political payoff lies in the power to block policies and programs, that is, in *negative* rather than positive authority. Even negative authority requires political access and support; thus the environmentalists were reticent to rank success in specific land use conflicts over the less tangible political and social accomplishments that accompany consensus building.

It may be too early to herald a transformation in democratic commitment, but the Age of Ecology has changed the way people think about information and expertise. Information now plays an unexpected role in environmental politics: not as a barrier to *participation*, as would have been expected in the centralist literature, but as a barrier to policy *implementation*.

This chapter reports on these trends in respondents' answers and attempts to explain similarities and differences between the two cases. Chapter 6 goes on to test the importance of the three challenges to democracy.

Participation and Control

Recall from chapter 2 that the centralists and decentralists disagree sharply over the environmental benefits of local political control. Decentralists predict that participation will increase local control over environmental outcomes: the more people participate, the more they will learn about issues and about the policy process. This prediction is essentially the premise of hypothesis 4, described in chapters 1 and 4, that people who participate intensively on environmental issues gain more control over environmental policy outcomes.

A related question is whether people actually *desire* more local control over the environmental policy process, or whether they have reasons to prefer more centralized authority. Centralists would argue that hypothesis 1 is probably true: if people perceive that they are in an environmental crisis, they will seek centralized leadership.

But contrary to the decentralist premise, I found that those who invested a great deal of time as activists and lobbyists at the community level did not characterize their participation as very effective at meeting their goals. For example, residents of Hayfork, in Trinity County, California, mounted a recall effort to remove two county supervisors they felt were hostile to the timber industry. They were successful in recalling one of the supervisors; the second survived the recall and went on to win reelection. Although their effort succeeded in sending a strong message and removing one supervisor from office, it was a hollow victory: approximately three-quarters of the land logged for timber in Trinity county is federally owned; thus, court-ordered protections for the northern spotted owl have a much greater effect on available timber for the mills of Trinity county than any action by the supervisors. One woman did emerge as a spokesperson for logging communities and in 1991 began to speak around the country on behalf of loggers, but Forest Service officials working in the Trinity National Forest did not believe that Hayfork's actions would significantly change Forest Service policies at the local level.

A similar story can be told of the neighborhood organizers in the "Victory Village" area of Sunnyvale, California (named for the war effort in 1943), who live near a Westinghouse site contaminated with PCBS. Neighborhood organizers wanted the site excavated to 50 feet—the maximum option considered by the EPA. When they did not succeed in obtaining that degree of remediation, they were shocked to learn that EPA could choose a "cost-effective" option instead of maximal cleanup. Despite the residents' presence at hearings, their letters to the EPA, and their growing influence with the city council and local legislators, the administering agency maintained its own prerogatives on the details of the cleanup decision.

In many disputes, control was exercised much more by the usual available avenues of opposition—namely, lawsuits (e.g., injunctions against harvesting) and legislation (Oregon's 1990 protimber Forest Practices Act)—than by direct participation. But participation in toxics controversies has become a widely (though sometimes grudgingly) accepted norm. Respondents in the toxics case placed a high premium on participation, with a growing stress on citizen oversight, vetoes, and management. It is now very difficult to seriously advocate anything less than frequent consultation with citizens in siting issues, and thus the rhetoric of industry, legislators, and agency staff—if not always their actions—was strongly proparticipation. John Rosell, a director of community and government relations for Chemical Waste Management said: "The best community advisory groups are the ones that have some say in some management decisions. [It's] best if it's done before the process is up and running."

What should control, or "some say in management," consist of? Elliott (1984) argued that successful community control over hazardous waste management facilities must have three components:

1. The assignment of contingent responsibilities
2. The design of a surveillance and feedback system
3. The structuring of an interactive decision-making process

Based on the 64 conditions of the facility's conditional use permit, Chemical Waste Management's desire to share some management decisions with citizens includes mostly condition number 2, surveillance and feedback systems (e.g., continuous emissions monitoring systems and computer links to agencies). To a lesser degree, Chem Waste agreed to demands for interactive management that would be characterized by condition 3. While residents are rarely interested in the details, responsibility, and commitment of micromanagement, they often express a desire to exercise control and oversight at critical junctures. Contingent responsibilities describe the actions facilities must take given objectively verifiable events (e.g., the facility closes if more than *x* number of delivery trucks arrive in one day, or process changes are instituted upon a majority recommendation by a local advisory committee). A few contingent responsibilities were spelled out in the conditional use permit, but they provided little, if any, role for residents living near the facility.

Spokespersons for Chem Waste considered the link between local control and environmental outcomes to be positive as long as communities do not interfere with facility siting or management after the facility is running. This view contrasts sharply with that of those activists living near such facilities in Livermore and Kettleman City: "We absolutely must have local control; we're demanding information on waste generators [and] a role, and trying to get the community a place at the table. [I] can't tell what the deficiencies of coequal membership might be, but [we] think the community should have an absolute say on incinerators, waste burial, and whether certain waste generating activity should be *here*."

Ultimately, the only link most industry respondents in the toxics case could definitively make between participation and environmental outcomes was simply that "the more participation there is, the less bad happens." Whether "less bad happening" is environmentally desirable depends on what the status quo is; more stockpiling and out-of-state transport of hazardous wastes may not be any more risky than a hazardous waste incinerator, and of course such risks are distributed very unevenly.

Forest preservation activists and members of timber associations alike felt that they had very little control over the policy process. Locally, environmentalists felt stymied by large timber companies, especially if they perceived that the logging industry would retaliate against criticism of harvest practices, for example, by harvesting disputed timber sales before logging could be delayed by court action. They also felt they were being unfairly treated by congressional committees, especially those that oversee Forest Service budgets and timber sales. Tim Hermach, director of the Native Forest Council in Eugene, Oregon, argued that most of the participation typically employed by environmental organizations succeeded only in wresting concessions from environmentalists or wasting their time: "What good is having access if you're compromising everything away? Throughout the entire history of the Sierra Club we've lost public land. They've never had an honest victory."

Hermach's opponents in the timber industry also felt that their participation in negotiations, hearings, and summits was never successful, if only because there was always some group "outside" of the process who could sabotage agreements. A timber industry member in northern California stressed that extremists on both sides made participation useless:

> At a pure policy level, you could probably reach consensus with people like the Audubon Society and Sierra Club versus the more enlightened fraction of the resource-using community. I think that is a real possibility. But what you would have to do to implement that strategy is change the laws so that the fringes on either side of the issue, whether it be . . . those who would still abuse the resource or those whose sole purpose for being was to stop all utilization of the resource, would not still have an opportunity to do so.

Moreover, the timber industry sees itself outspent and outlobbied by environmentalists. The vice president of a local bank in northern California underlined this feeling of helplessness: "When I was in the timber industry, people really [had] a mistaken notion of power. Where is this power? As soon as [some issue] was over, we'd go back to fighting among ourselves. . . . people talk about the forest products industry as being some sort of monolith that walks in lockstep. Nothing could be farther from the truth." The only respondents in the forestry case who thought that they had been effective were either the few environmentalists who had won major legal challenges or the timber industry lobbyists. The latter succeeded in averting a major Oregon forestry initiative in 1990 by garnering enough votes in the Oregon state legislature to pass a protimber reform bill. Neither strategy relied on participatory democracy.

The Locus of Control and Environmental Outcomes

Another difference between the centralists and decentralists turns on the desirability (for environmental purposes) of local control. The centralists' core premise was that an ecological Leviathan, freed from the fetters of foolish, tragic democrats, would act decisively to save the planet. The decentralists countered with the premise of hypothesis 2: local control over environmental management results in more desirable environmental outcomes than central control (e.g., state, federal, or frankly authoritarian).

Despite their past experiences, people involved in the two cases of this study believe that successful, local environmental policy control is desirable but rarely a real option. Activists and industry personnel alike must constantly assess the effectiveness of their existing political strategies versus what might be accomplished if they could change the venue of their influence. Most respondents had some unspoken causal theory in mind about the environmental outcomes that they could expect if forest resources or hazardous wastes were locally instead of centrally managed. But we can demonstrate, in theory and practice, that there is no logical reason why sustainable and desirable environmental outcomes should proceed from either a decentralized or centralized mode of control; indeed, either form of governance can result in a positive or disastrous environmental outcome. Thus, it was crucial to ask respondents about their causal assumptions not so much because we then might learn who is "right" about the "best" locus of control for environmental sustainability, but because their beliefs affect whether they will view decentralized or centralized systems favorably.

After asking respondents whether they favored local control, I asked whether people living near the national forest would manage it better than remote agencies, or if the waste problem would have been this bad if respondents had had more input. *Generally, environmental activists, legislators, and legislative staff in the toxics debate assumed that increased local participation and control would result in better environmental outcomes but held the opposite view on forestry.* One resident of Kettleman City opposing Chem Waste Management's incinerator offered an unsolicited, succinct assessment: "More participation equals less waste."

But agency officials in the state Department of Toxic Substances Control Division (DTSC) were less sanguine about local oversight of hazardous waste treatment, fearing that California would never achieve proper treatment capacity levels if the state could not impose more facilities on local areas. One DTSC official argued that locals would refuse to accept their fair share of the state's hazardous waste treatment capacity because they are unwilling to see the trade-offs inherent in an industrial society, namely that we can't refine oil without creating wastes and that source reduction will never eliminate hazardous wastes altogether.

Legislative and agency staff may assign different values to participation and actual treatment efforts because of their professional commitments. Although both spend a great deal of time reviewing technical documents and proposals, legislative staff tended to be more concerned with their legislator's constituents and expressed concern over citizens' political rights and questions of "empowerment." Agency staff were more focused on moving the state toward self-sufficiency in hazardous waste treatment, lest California become a pariah among the states accepting its wastes. Finally, agency officials expressed concern that local control or oversight in hazardous waste management was almost always construed to mean local vetoes, or in fact "no facilities sited at all." And agency officials feared that if only large companies had the resources and willpower to stick out a protracted fight over siting, a handful of monoliths would be the only industry players in California toxics management.[3]

Also worried about fueling polarization among their constituents, legislative staff argued that members of the legislature could not afford strictly parochial views. Compromise was the theme of Rick Dunne's[4] opinion about local control:

> The problem that you've got is that the answer I'm going to give you sounds like hedging. There is no answer. You've got to have all of the component parts. You've got to have the local participation, because it's their economy and environment that are directly impacted; you have to involve the larger bureaucracy and the different entities overseeing the good of the entire ecosystem. For example, let's [say that] in Fort Bragg they wanted to come out and govern how they harvest trees, they're more interested—because of the employment picture—in cutting trees than in preserving the ecosystem. Or in doing sustained yield. They just want to get in there, cut, and get out. . . . what has to be done, and is certainly not being done—except in isolated pockets like Mendocino County—is more local participation where [people] can get together with

the leadership and decide on a course of action, and take that course of action out to the state and federal level.

For Dunne, a balance of power is necessary because increased participation is not likely to result in more "rational" management:

The community is going to have a very selfish position. They're either going to perceive it as "I've got a wife and two children and I'm thirty years old, and I've been working in this lumber mill for two years and now they're going to shut it down, what the hell do I care about the spotted owl, chop it down." Is that a rational decision? No. On the other side of the coin, "I'm sixty-five years old, I'm retired, I moved up here because there are plenty of trees and the town is quiet, and now they're going to cut them all down, save the trees." Is that rational? Not really, either. There's a logical approach of doing it, but you've got to have input from both of those communities before you can make a logical decision on how to best manage the resource.

Loggers and protimber industry mostly favored local control, and Forest Service staff thought that strong local influence (as opposed to de jure control) would be mostly benign if tempered by the Service's control over final decisions. The general manager of a biomass waste-to-energy plant in northern California likened the state's regional antagonisms to the breakup of the Soviet Union:

You could apply the same logic to the northern part of California wanting to be their own state. That we can be, we know we can be far more efficient, we know we can make more correct decisions regarding our resources, because we're closer to them. Just because we become our own state, as has now been proposed, that doesn't mean it's because we want to damage them [resources], it means we want to have the decision making power to utilize those resources . . . properly. And not be hamstrung by a group of 20 million people south of us that have no knowledge of those resources. . . . There's a certain attractiveness to that. Why would you foul your own nest? If you want to leave it at least as good as you found it—all of us want to do that.

Some protimber respondents expressed a libertarian belief that people using resources know what's best for them and ought to be left alone to manage them as they see fit, within only the broadest state or federal constraints. A member of a proforest products industry group in one of the inland valleys of northern California insisted that conditions changed too much from region to region, and thus a statewide, brokered approach to forest management was unworkable:

> We're making statewide decisions. There are very few decisions left to the local area. And it's really sinister with the Forest Service. Their decisions are essentially coming from Washington, D.C. And most of them are PCT—politically correct thinking. Trying to outguess the perception of the public, and trying to do what the public thinks is correct. And to a lesser degree you've got the same problem with the state, and then their impact on private lands. . . . I can't judge what should happen in Eldorado County. I don't know all the plant species, I don't know their relationships. What they have to do to handle reforestation . . . has got to be a different way of doing business. That opportunity is taken away by state law.

Forest Service and California Department of Forestry and Fire Protection (CDF) staff were more cautiously optimistic about the effects of local control. Supervisors in Plumas and Siskiyou counties thought that local residents, most of whom had strong protimber attitudes, would probably like to see the currently mandated allowable sale quantity (ASQ) go up, but "definitely would not want to overcut." Over and over, foresters like John Kruse, a forest planner in the Shasta-Trinity National Forest, stressed that local communities, by and large, would be capable and responsible enough to seek balance, especially if they understood the multiple-use constraints of the Forest Service and worked with foresters: "If local folks, including some of the environmental groups, had control [over national forest management], the pendulum wouldn't swing as wide."

Of course, one's definition of "multiple-use" helps determine what is considered a desirable outcome of local influence. For example, foresters and timber industry supporters can reconcile themselves to "habitat protection" as a goal of multiple-use if very specific objectives in, say, old-growth preservation, have been identified: "If you cut trees of all age groups, you could [still] maintain the attributes of old-growth necessary for the spotted owl, like nesting trees, foraging habitat, dead and down trees." This is not old-growth preservation per se, and many ecologists would think of the biological benefits of old-growth as far more than just spotted owl habitat. But the more a popular ecological prescription (e.g., "save the old-growth") can be simplified, the greater the polarization of the old-growth dilemma into a conflict between only the owl and timber industries. In turn, antagonists tend to caricature the preservationist task of multiple-use management.

These observations contrast sharply with environmentalist respondents' assumptions about local control and environmental outcomes. For example, some environmentalists in California were strongly supportive of decentralizing influence over forest land use decisions by increasing local sawmill ownership. They were convinced that timber industry labor and environmentalists were natural allies, on the theory that large companies were responsible for "boom and bust" cycles in the timber market. These cycles often lead to unstable employment and rapid liquidation of high-value timber. For those environmentalists focused on labor and mills, local control has the potential of being more environmentally benign than either private ownership of mills and timberlands by large, interstate companies or state and federal management.

Many more environmentalists, however, based their assessment of the environmental impacts of local influence and control on a more stereotypical depiction of timber executives and loggers. Tim Hermach, of the Native Forest Council, was horrified at the thought of local control: "Local communities should *not* control local forests. . . . you'll get some local bigot fostering inequalities, [because] it's more plausible to fool some of the people some of the time in *smaller* places."

The environmentalists in Oregon's Cascades and in California's North Coast were more likely to adopt the rather elitist view that loggers and community residents were duped or held hostage by the dominant timber companies in their areas. They repeated many times that the timber companies would harvest their way out of any chance at maintaining a sustainable timber economy and that loggers excessively discount the future. But environmentalists have had a hard time convincing labor of their analysis. Concerning a labor-environment alliance, Andy Kerr, the director of the Oregon Natural Resources Council (ONRC), said:

> It would be helpful [to have the support of labor], but not required. It is mythology and a polyannish view to say that environment and labor are natural partners. . . . most loggers have very short-term time horizons. You could say [to the loggers], "You're going to lose your job [because of

overcutting] in five years." That's 60 months! They could be unemployed three times in that period. It's one of the last parts of the American economy that provides relatively high-paying jobs to uneducated people.

The conviction that locals are not to be trusted with forest management has led many environmentalists to embrace a national, and largely urban, constituency. Although they have little success controlling what happens on private lands, their rhetoric repeatedly stresses the *public* ownership of national forests and cites provisions of national legislation that support their positions. For environmentalists seeking to protect Opal Creek and Mt. Hood in Oregon or the Headwaters Redwoods in California the battle is largely national. Michael Jones, of the Cascade Holistic Geographic Society near Mt. Hood, "always [tries] to stress that this area doesn't belong to the people who live here." Jones might not have appealed to such a broad constituency if he felt committed to local authority. But he was politically pragmatic, like most of the environmentalists I interviewed.

Even a national battle relying on mass appeal is problematic for environmentalists, as this comment from Dave Foreman, cofounder of Earth First! attests:

> I am terrified of mobs, I am scared of mass movements. I'm very frightened of organized human society, particularly official ones we call "the authority." . . . I guess I'm enough of a Jeffersonian that I really believe in individual action. . . . I think when people do things by themselves thoughtfully that they're a lot safer. I think when you get a mass movement together, things happen psychologically. And people who are leaders of mass movements—and there are always going to be leaders—whether official or unofficial, have an enormous responsibility. But we do need mass movements. . . . I would love to see half a million people march on Washington, D.C. demanding an end to cutting in our national forests. We need that. I don't see it happening. And if we could do it, I would be a little bit nervous, even if I was at the head of that march. I would be very nervous. I would be nervous about what it was doing to me, and I would be nervous about what it was doing to the people in that mass movement. And so, it's just a caution I'm throwing out there.

How can we account for the differences across cases and respondent groups in how local control and environmental outcomes are thought to be linked? One possibility is that an urban-rural split goes a long way in defining the positions of opponents. Both environmentalists and their opponents agree that conflicts are exacerbated by differing perceptions in urban and rural areas. In the case of toxics, waste sites exist in so many communities around the country that for every one of them affected by hazardous waste dilemmas, the issue is strongly local as well as strongly national. This is not so true of forests and has led to deep resentment among locals who fear that urban populations cannot grant them a fair hearing; as this northern California bank executive put it:

> The impression of the forest through the media, removal of people from the land, the fact that their personal environment stinks; the best place to do something good for the environment is where somebody else lives. I've convinced myself that I could stand up to any group and say that the day development is stopped in San Diego, Orange and Riverside counties to save the gnatcatcher, that they stop dredging San Francisco Bay to protect winter-run salmon, the day that they stop providing 85 percent of CA's water . . . by ceasing pumping from the Delta to protect the Delta smelt is the day I'll accept *anything* they want to do to protect the Spotted Owl. Do those three things for me. . . . yeah they'll be some dislocation, doesn't bother me one bit.

The tendency of environmentalists to use most means available to achieve physical outcomes may explain why the environmentalists in the toxics cases and those in the forestry cases viewed the effects of local control so differently. People who have never been "active" on anything but environmental problems may exhibit a single-issue focus; they may pragmatically adjust their strategies in whatever way necessary to achieve their objectives, and if that means appealing to a national constituency on one issue but not another, no ideological or moral commitment will have been breached.

In the toxics case, local environmentalists cultivate NIMBYism and use it to score significant, sustained victories. In the forest case, the environmentalists are defeated by a simple numbers game: more antitimber support comes from urban areas distant from sites of controversy than from rural communities whose economies are tied to resource extraction. Mobilization in the forestry case thus necessarily requires activists to "educate" their target audiences in a way different from that of the antitoxics groups who can use fear as a potent tool for garnering support. The implication from these two cases is that the locus of decision making is an instrumental, not a philosophical, consideration for most activists.

Moreover, in thinking about the human health dimension of the toxics debacle, the toxics activists assume—often correctly—that health hazards are something the public quickly grasps as a key issue about hazardous waste. As long as toxics remain a human health concern, environmentalists feel sure that, once they are minimally informed, locally affected citizens will opt for safer hazardous waste management, or simply relocation of the hazardous waste problem away from their communities.

Though environmentalists generally thought they had public support, participants in the two cases saw their relationship to the "mass public" in different ways. For members of the forest products industry, the public has not been on their side, and several respondents from this group considered the public to be out of touch with resource industries. Thus, loggers did not consider the public a natural ally so much as a wayward relative who used to understand the values of hardworking outdoorsmen or who pictured lumberjacks to be folk heroes like Paul Bunyan.

Industry respondents in the toxics case made more of their relationship to the public, albeit guardedly. That may be explained by the nature of the NIMBY syndrome. In the forestry case, environmental activists argue that all Americans are aggrieved by overcutting, but in the toxics case, the dangers of hazardous wastes are very localized. Thus industry officials can safely assume that most Americans do not feel personally and immediately threatened by toxics. Arguably, activists need to make less effort at overcoming a negative stigma to gain public support—at least for the *idea* of effective hazardous waste management.

Environmentalists' relationship with the public has been generally quite good,[5] but environmentalists must continually combat the perception that they are not all that interested in people (which encourages the counter-assessment from the public that it is sometimes not all that interested in environmentalists!). Their ability to reverse this perception is the reason that neighborhood activists in the toxics case can "deliver" opposing votes on toxics issues more consistently than can environmentalists in the forest case. The toxics activists will always have the upper hand as long as they can frame the debate in terms of health and keep it away from a discussion of competing economic scenarios or political philosophies.

"Success": Political Procedures or Physical Outcomes?

If people are politically pragmatic about whose participation is best for the environment, what are their ultimate goals? Will they rank some political ends over others? An important claim of the centralists is that, in an environmental crisis, people will become desperate enough to accept a draconian solution—even if it costs them their rights of self-determination. In essence, the theory goes, people will rank environmental protection over political structures that cannot seem to deliver ecological solutions, and the "fairness" of political processes will come to be judged ex post facto by the environmental outcomes they achieve. Stated differently, the centralists warn that "environmental justice" will subsume the entire meaning of social justice, leaving no room for considering *which* political path achieves environmental protection. This is the message of hypothesis 3: people who participate in the policy process over environmental dilemmas judge the value of their experience by what happens in the physical environment.

The decentralists suggest otherwise. They believe that environmental justice *is* social justice but not *only* because the right outcomes are achieved. Environmental justice and social justice cannot be separated because neither precedes the other, and participatory democracy is an integral part of the social and environmental definitions of justice. Thus, the decentralists portray a two-way causal link between democratic politics and environmental sustainability.

Do the empirical responses suggest that people have procedural goals in mind when they participate, or are such aims contained in strategic, substantive goals? What role does the notion of community play in respondents' actions and in their definition of success? Is community building a great enough value to motivate participation? As mentioned earlier, asking how people define success provides a way to elicit whether respondents are more concerned about procedural goals like participation or about achieving a particular state of affairs. The important question here is, When forced to make a choice, would the respondents sacrifice their preferred environmental objectives in favor of maintaining a minimum acceptable level of democratic participation? Did they, in fact, have cause to view the two as competing values?

Most respondents were not willing to rank participation and environmen-

tal protection or environmental management goals, though these principles often competed during interviews. Some, especially in the timber and waste management industries, did express weariness with the process they had to go through to oppose or promote some action that they felt got in the way of "just doing my job." Such frustration notwithstanding, most people did not separate the participation they were entitled to from environmental goals; in fact, a loss in an environmental battle was often linked in people's minds with "not enough participation." As Luke Cole, a lawyer for California Rural Legal Assistance (CRLA), said: "Success is putting in place a structure that empowers people to take charge over their lives." Because hazardous waste issues are conducive to local involvement, it is not surprising that he assumed negative environmental conditions to prevail where people lacked "empowerment," and that he was convinced that more participation would result in better waste management. Thus, by necessity his efforts were focused on strengthening the ability of local groups to make their voices heard.

Certainly all respondents measured success partly by how well they thought their messages were getting across, but some were more concerned than others with specific goals. In fact, "success" was defined as one or more of three possible outcomes. Respondents emphasized (1) primarily political or procedural goals, (2) specific physical outcomes, or (3) economic objectives for themselves and their communities.

Sociopolitical Success

Respondents offered two kinds of social and political versions of success. The first was simply achieving more access to the political process—becoming a regular participant at hearings or on joint projects (especially with the natural resource agencies); electing "one's own" to school boards, the county board of supervisors, or city councils; forming environmental planning commissions; and increasing the influence of local trade associations. Success for forest product industry members in Shasta and Trinity counties meant that they could have the Forest Service's ear through their Shasta Alliance for Resources and Environment (SHARE). SHARE succeeded most when its criticisms of Forest Service timber harvest plans motivated the Service to revise entire planning documents.

Local elections spelled success for some environmentalists. Activists took credit for electing Hispanics to the school board and water board in Kings County, which is the site of Chemical Waste Management's hazardous waste landfill. And while Anna-Marie Stenberg, a member of Earth First! and of the International Woodworkers Union (IWW), made an unsuccessful bid for Mendocino County's Fourth District supervisor seat, residents of Hayfork (Trinity County, California) recalled a supervisor they felt as not protimber, as discussed earlier. Bob Fredenberg, principal consultant to the California Senate Toxics Committee summed up the view of success was political access by saying: "Making people a political force will achieve these policies [Superfund site cleanups and industry-wide pollution prevention]. Any impacted community will want less waste. . . . even nonimpacted communities can see that it could or does happen in their communities; just look at the medfly spraying."

A second social and political version of success centered around building consensus and community. In this case, respondents were not just interested in gaining access to decision making but expressed a hope that their activities would help heal the divisions and conflicts in their communities. One neighborhood activist in Sunnyvale said that even if she did not get the kind of site cleanup she wanted, she "now knows [her] neighbors in a six-block radius." Many respondents wanted to tone down the polarization and conflict in their communities. As a member of the Yellow Ribbon Coalition, a loggers' advocacy group in Springfield, Oregon, said: "[I'd like to see] balance, communities working, maintaining [Oregon as a] natural resource state, a multiple-use state. I'd like to see environmentalism stop being a money industry; more grass-roots, and less conflict. . . . no conflict would be best."

Naomi Wagner, an environmental activist in Mendocino County, California hoped that the Mendocino County Board of Supervisors would pass a bold set of "special county rules" to control logging practices on private land. The Board failed to pass the special rules by a three to two vote, but Wagner felt that the whole effort had led to a change in rhetoric and the "bottom line." In her view there was value in the fact that Mendocino County had empanelled a forest advisory committee to recommend "sustainable" practices (i.e., the "special rules") and that the word "sustainable" was now on everybody's lips. As Wagner put it:

> The basis of the discussion has changed; we're not talking about owls and daddy's job, because everyone is unemployed. People "get it." . . .

Big timber doesn't dare pull that story. There's no trees left. . . . L-P [Louisiana-Pacific] and G-P [Georgia-Pacific] are filing very few timber plans in this county. . . . L-P has one mill with one shift going.

For Wagner, "big timber" was out of the picture because Mendocino County had little commercial timber left to cut. Now Wagner and her colleagues have joined with unemployed timber workers to practice "sustainable" forestry in a new labor-environment coalition called the Mendocino Real Wood Co-Op. And Mendocino County is not the first to try forging coalitions between traditional antagonists. Jim Workman directs the Rogue Institute for Ecology and Economy, a labor and environment coalition in southern Oregon. He felt that loggers, millworkers, and environmentalists should break down old animosities imposed upon them by economic forces outside of their communities: "Success would be a lot more small manufacturing facilities throughout the area; [and a] pride of place. . . . it's *my* forest, [this is] what *we* did to manage it."

Rhonda Rigenhagen, a public information official at ROMIC (a private hazardous waste recycler) echoed Wagner's and Workman's sentiments, saying that "success is when there's agreement on all sides, and finding hazardous waste management solutions everyone can live with." Rigenhagen, members of the Yellow Ribbon Coalition, and other respondents were aware of and concerned about the viability of deeply polarized communities. Loggers and rural environmentalists deeply regretted the strife and anger surrounding them, while waste management proponents wanted a community to accept their presence. Dow Chemical's Bryant Fischback[6] genuinely regretted his community's opposition to a proposed on-site hazardous waste incinerator: "[I] would like to see community advisory panels be very straightforward and constructively critical . . . [and for Dow] to become increasingly responsive and ultimately viewed as a highly responsible neighbor . . . [and to] respond to needs and concerns of the community to stay here a long time."

Jim Workman's vision bridges that of the environmentalists, who generally sought the first definition of political success, that of gaining access, and that of proindustry (chemical, waste management, and timber) respondents, who were more concerned with building community, the second definition of success. Perhaps this subtle distinction helps explain why environmentalists appear to be less concerned with people than with nature. Environmentalist respondents were indeed concerned with building community; they simply defined their community to include trees and wildlife on a par with people.

Agency staff tended not to describe any political objectives in their definitions of success; they had more strictly environmental goals in mind. One high-ranking DTSC official proved the exception in arguing that toxics education had to be linked to lifestyle choices. The degree to which that linkage could be effectively communicated, and to which it altered people's behaviors, constituted a critical component of his definition of success:

> Success for me would be educating people in terms of what the costs are of lifestyle choices in a fashion that they [people] (1) be responsible for their personal lifestyle, and that's everything from buying a nylon rope to driving your car; [and] (2) we would have educated people to the point where they realize that they have to be responsible for their lifestyle choices in terms of waste management. . . . we need to have a dramatic shift in the way people relate to toxics issues. Right now people generally want to be able to buy anything they want, go wherever they want, in their own car, and they don't want anything related to waste management anywhere in their local environment. They're totally oblivious to reality.

Success "On the Ground"

Many environmentalists and neighborhood activists wanted nothing less than substantial changes in the physical environment, even if they had to rely on "power brokers" or the courts to get their way. Instead of "more democracy," these respondents were looking for very specific physical outcomes: a site cleanup, a protected stand of trees, reduced or increased timber harvesting, or expedited hazardous waste treatment facility siting. An obvious example was the "Albion uprising," a direct-action protest that began in May of 1992. During that protest, environmentalists and residents of Albion (Mendocino County, California) "sat" in trees while litigants sparred over a timber harvest plan to cut 280 acres of second-growth redwoods. The environmentalists eventually prevailed through the courts.

Gordon Smith is the director of the Olympic Natural Resources Center, a clearinghouse and policy institute for Washington state natural resources, created by the state legislature and housed at the University of Washington. His version of success requires

> changing how our American forest resources on public and private lands are managed to better provide for biological diversity, maintaining high levels of primary biological productivity, maintaining ecosystem resilience, and maintaining a range of outputs desired by the people—that includes timber, fish, clean water, clean air, and recreation. That's sustainability—I would allow for some trade-offs: "Sometimes I'm going to do X damage, sometimes protect ecosystems."

Smith's comments were echoed over and over: George Atiyeh, a miner fighting to preserve an old forest near Oregon's Opal Creek, did not think that all logging on public lands must stop but wanted to see no more logging of old-growth; Andy Kerr of the ONRC wanted to see "change on the ground, and successfully enforced injunctions;" and a neighborhood activist with Tri-Valley Cares, a watchdog group near the Lawrence Livermore Laboratory, wanted it all:

> [Success would be] dismantling an incinerator, conversion of Lawrence Livermore Labs away from nuclear weapons research to socially just, environmentally sound research; (1) global nuclear disarmament, (2) demilitarization, (3) equitable distribution of wealth among nations, (4) action for ecological balance, (5) getting more people to speak out.

Legislative and agency staff also had visions of what should happen to the natural environment, but once again these respondents were more conciliatory toward competing interests. If, in the future, the DTSC official quoted earlier were to look back, what would he call success? He would want to see "self-sufficiency" in California's hazardous waste management, and "teamwork":

> A way to identify [what] would be successful is if we're *not* shipping all our waste out of state, and that we are responsibly accepting waste management facilities in proximity to waste generation, and that we've reached the point where environmentalists . . . are interacting with business [people] in the context of [a] societal teamwork approach to environmental problem solving.

Economic Success

Centralists and decentralists generally do agree on one claim. Both believe that economic growth, as it has been pursued and practiced, is largely responsible for environmental degradation. In addition to its environmental effects, theorists have argued that economic growth has taken on such a compelling imperative of its own that it is politically taboo to suggest curtailing or redefining it—even to reverse environmental blight. In effect, centralists and decentralists agree with hypothesis 6: concerns over economic growth sharply constrain the scope and number of distinct environmental policy options that are seriously considered by participants in the environmental policy process.

Where respondents gave economic growth and well-being as the primary definition of success, it is reasonable to infer that they felt obliged to discount policy options and proposals that they perceived as subordinating economic growth. Following this theme, respondents were asked whether they could or should consider policy options with substantial environmental gains but potential economic costs.

Not surprisingly, economic successes were most often the concern of timber industry officials, loggers, and chemical or waste management industry spokespersons. They wanted to maintain timber jobs, gain support for economic transition programs, and treat hazardous wastes in a cost-effective manner. For these respondents, success is a matter of economic survival. The owner of a small timber company working the Shasta-Trinity area said:

> The average logger is sitting in close to three-quarters of a million dollars in debt for equipment . . . and his hope is that he will have a reasonably decent life, pay his bills, support his crew and his family. . . . he might sell his equipment when he gets ready to retire, and that will be his retirement. Very few of them will ever have a lot of money. . . . you deal with nature every day; nature throws you enough curves, and now we've added additional curves with all this up and down of the political side of things. . . . even fewer will end up with anything to sell.

Gordon Smith, of the Olympic Natural Resources Center, suggested one reason that loggers may define success in economic terms, aside from the obvious fact that their livelihood depends on it:

> There are a lot of myths about what is just in the world. It's a lot more respectable to a logger if he loses a job because there are no more trees to cut, or because the economy's gone to hell and it [the economy] just does that, than it is if enough of the 300 million people [*sic*] in this country say those trees are worth more to us standing than turned into lumber. It's a proprietary feeling . . . "you made a promise that I'd keep having this job." It's also partly [a feeling] "what the hell are you doing taking my job away?" And partly [a belief that] the ecosystem is fine without preservation.

Sometimes economic successes may occur even where physical outcomes have not been achieved. Kip Lipper, California assemblyman Sher's chief toxics staffperson, drew a parallel between the antinuclear power movement in the 1970s and environmental outcomes of NIMBYism in California. He pointed out that the antinuclear protests contributed to energy conservation in California.[7] "Similarly, the pressure to block some of these [hazardous waste treatment] sites has promoted source reduction. . . . those are good things. . . . it would be presumptuous of me to question democracy in action, but it [NIMBY] has and will change the economics of disposal." Thus, economic success can also be measured in terms of how much the current waste stream has been reduced (thereby lowering disposal costs), even if no new treatment capacity is available to treat the oversupply of existing wastes.

The environmentalists' pragmatism—so prevalent in their assumptions about the environmental benefits of local control—was somewhat tempered in their definitions of success. While their attention was certainly fixed on "the environmental prize," environmentalists also recognized that there were psychological benefits to participating and becoming activists. From a motivational point of view, this is a good perspective to have if you or your group more often loses than wins; intangible rewards like "community building" can offset or compensate somewhat for disappointing and lengthy struggles.

Conflict resolution and community building were also important to people in logging areas, who were there precisely because they wished to preserve shared values that brought them—or kept them—in the countryside to live and work. These sentiments may have come from a view that environmentalists were outsiders who really shouldn't have the influence they do, and who lacked the legitimacy even to seek control over the lives of people working in rural communities. And as Gordon Smith pointed out, the loggers' and timber industry's first instincts tell them that environmental activism is equal to illicit tampering with "the free market" and that any political claim that can affect timber jobs so much should really not be open to discussion.

Generalizations must be tentative when based on only fifty or so in-depth conversations, even if the views of the interviewees are well known. Still, the question of success did reveal some important points about the principles of environmental sustainability and participatory democracy. Among environmentalists and loggers/industrialists, it appeared that the value of political participation was somewhat subordinate to other principles. In the case of environmentalists, procedural access was important, but most were narrowly focused on the environmental, physical goals they mobilized to achieve. Loggers and industrialists viewed participation as a necessary hurdle to getting on with their work. With a few exceptions, participation was most often of instrumental value and rarely prized for its own sake. This by no means suggests that participation is expendable, but rather that many define it simply as a means to another end whose value may change as more important goals or principles evolve.

Technical Information and Citizen Competence

Recall from chapter 3 that technical information might constitute an important part of the technocratic challenge to democracy if indeed it hampered people's participation in environmental policy disputes. In fact, most centralists stress that a fundamental problem with the decentralist prescription is that it exaggerates people's ability to make meaningful and reasoned decisions on technically complex matters. The decentralists argue that the centralists have underplayed the sinister, secretive tendencies of a ruling cadre of technocratic elites. The decentralists thus advance hypothesis 5: technical complexity is used by experts to block or negate public participation on environmental issues. Similarly, according to this view, nonexperts affected by environmental controversies limit their own participation because technical complexity intimidates them and prevents them from identifying and effectively advancing their goals.

This section summarizes the respondents' answers to questions concerning technical complexity and participation. Contrary to what might be expected if technocracies manipulated information to maintain or enhance their own power, for most respondents information was not an issue. For example, no one seriously suggested that technocrats (e.g., people referred to as government scientists, engineers, or other civil servants) should be left to manage the two issues studied without some profound and strong guidance from both the public and elected representatives, or that technocrats should have more discretion because they "knew better" or had access to "better information." Respondents often insisted that such guidance had to be in the form of clear statements of goals and objectives: for example, "The purpose of forest management is to provide timber and employment," or, "Clean water has no manmade chemicals in it." Participants wanted objectives to be both detailed and adaptable to many situations, and they were not willing to leave the determination of these goals to technocrats.

Expertise wasn't a barrier to participation for members of organizations like SHARE, because they were technically articulate. They could speak competently on forestry and habitat management and on environmental issues completely unrelated to their everyday experiences, like acid rain and global warming. SHARE members also had a good grasp of how the resource agencies operated, what pressures they faced, and what requirements they had to meet. Environmentalists, too, were well-informed. They not only understood what information they had to have to discuss a site cleanup plan or an environmental impact report but also knew where to go for information. Most important, respondents on all sides of the issues in both cases recognized that not everyone needed to be highly expert in order to participate effectively.

Ten to fifteen years ago, when the term NIMBY was coined, technical complexity was considered a more intimidating barrier to effective participation (Morell, 1992). Today, neighborhood activists have learned how to seek technical help, and an entire community of volunteers and for-profit consultants is now at these groups' disposal. Even the federal government's aid has enhanced citizen competence through the use of technical assistance grants (TAG) under Superfund, to allow local communities to independently seek out their own expertise. Advocacy coalitions have also increased their rate of policy learning, thereby shortening the time it takes for participants to become technically and politically competent.

Interestingly, such learning was not confined to the seasoned neighborhood organizers in the toxics case. Sustained conflicts over hazardous waste management have evolved through more phases than forest disputes have, but activists in numerous environmental disputes other than toxics have learned to quickly become citizen experts. Increased issue knowledge has been characteristic of disputes over nuclear waste and energy and large water projects, to name but a few (Kraft and Clary, 1991; Mazmanian and Nienaber, 1979).

Increased familiarity with policy issues implies that it may take less time for potential participants to acquire and use information. But if technical knowledge is no longer a barrier to participation, it is still no less of a concern and source of distrust for participants in environmental controversies. Information as a political tool has taken its next logical evolutionary step: even if everyone can obtain *enough* technical information—so that it can no longer be invoked as a vague but exclusionary pedigree—participants, despite shared sources of information, will still arrive at dramatically different conclusions. Ironically, the same technical information and expertise that gave rise to the Age of Ecology is being treated as a political football. And since the framing of the "right" questions for science to ask (e.g., Can spotted owls survive in non-old-growth stands?), or of the right goals to articulate (Does it matter if spotted owls survive in non-old-growth stands?), tends to determine the direction of political debate, the temptation to devalue all information (and its interpretation) but one's own becomes very great.

Devalued information also becomes useless for policy-making, reinforcing the political impasse of environmental policy in places like northern California. Moreover, ardent calls for "objective centers of information" where the scientific merits of different observations and policy recommendations can be "reasonably" and "rationally" evaluated are probably naïve or disingenuous. Many respondents in the forest case expressed a belief that "objective" information would support their views, and that the operative problem consisted of gathering and disseminating "honest" research. But foresters and environmentalists, loggers and county supervisors all see what they want to see when they go walking in the forest. In effect, the acrimony of the forestry debate in the Pacific Northwest may have turned knowledge and information about forest ecology into a social and political artifact, unable to stand on its own merits outside of the political arena.

Politicizing technical information worries those concerned with effective policy implementation. When everyone's facts, figures, and reports are suspect, exasperated antagonists tend to scoff at the "promise" of participation, claiming that it just slows down problem solving. And scientists and technocrats are frustrated no end, because they believe that they can answer questions about their objects of study and act on them in effective, rational ways. However, their expertise is routinely discounted in political exchanges. The new role of information also suggests that, in addition to the question What is possible? participants are concerned with What is desirable? To the extent that opponents in forest and hazardous waste management can argue on the level of goals and trade-offs, their debates may avoid the false pretense of relying on science and technical information for definitive answers.

The Promise of Regionalism

The different stages of awareness, conflict resolution, and alternative policy strategies that have characterized toxics cases over the last fifteen years suggest that hazardous waste policy has had more time to evolve as an issue of contention than has the forestry issue. While hazardous waste policy experiments have been attempted since the early 1980s, policy innovation in the forestry case has been attempted only more recently. Two aspects of toxics policy experiments in California have centered on citizen oversight and regional government. Because of NIMBYism, sponsors of hazardous waste treatment facilities have become more and more receptive to citizen advisory committees that are actually vested with a degree of management discretion over treatment facilities. At the same time, the concept of "fair share" has made its way through hazardous waste commissions and legislative bodies at the city, county, and state level. In fact, much of the reasoning behind the Tanner planning process (see chapter 1) stresses a fair share concept, whereby counties are expected to develop waste treatment capacity for the toxics produced within their own borders. Yet all counties might not need the same number of facilities handling all kinds of wastes, even if they produce similar amounts, and so regional compacts have been endorsed by municipalities, the state, and regional advisory bodies.

Among the most important regional planning entities in the state are the Southern California Hazardous Waste Management Authority (SCHWMA) and the Contra Costa County Hazardous Waste Task Force

(HWTF). SCHWMA and the Contra Costa County HWTF were formed to bring legislators, utilities, manufacturers, environmentalists, and agency staff together to discuss and address waste management. These policy experiments are well described elsewhere;[8] their importance for this study is that they represent a nested approach of direct participation within a regional governance setting, which has evolved as a way to address the gridlock and NIMBYism that still characterize the politics of hazardous wastes.

Although these approaches may have succeeded in bringing "diverse interests . . . together in an enduring, viable, region-wide working body,"[9] the level of treatment capacity in the state has still steadily declined with the closure of the Santa Barbara area Casmalia landfill and BKK Class I landfill in West Covina. In Contra Costa County, some small treatment facilities may operate in the future, but they will not be sized to accept wastes from multiple counties or from the whole state.

Given the contentiousness of facility siting in the 1970s and 1980s, Tanner planning should have been a tremendous policy breakthrough and was heralded as such in the toxics policy community at its introduction in 1986. The Tanner process was successful in redefining the toxics issue with a new focus on source reduction; however, if read as a step toward facility siting, the act has done little more than fan the fires of controversy since it was signed.

If broaching the difficult issues of hazardous waste treatment capacity is a first step, then regional groups like the SCHWMA, as well as the legislature (in the case of Tanner planning), have taken it. But as difficult as the first step was, the next phase—developing regional waste management—has not even begun:

> As we noted, if the standard of success is the actual siting of new landfills, treatment and residuals facilities, and substantial source reduction of hazardous waste, the record of accomplishment to date is not impressive. "Fair share" rules, the concepts of channeled liability for facility operators, compensation for host communities, and siting criteria have all been explored, and a much better picture of the entire hazardous waste and toxics picture has emerged. Some headway has even been made in the actual siting and development of one regional industrial park and waste treatment and transfer facility at the old Kaiser Fontana steel mill location. Each of these developments should ultimately contribute to the solution

of the hazardous waste problems of the region, but, as of yet, they remain promising ideas and policy objectives, not realities. (Mazmanian, Stanley-Jones, and Coreen, 1988, p. 64)

The Tanner planning process has had mixed success at developing partnerships between counties, regions, and state agencies. In 1988, all 58 counties in California went to great pains to develop mandated County Hazardous Waste Management Plans ("Tanner plans") outlining their treatment capacity needs and the all-important siting criteria they would impose on any new facility proposals. But the state agency then responsible for overseeing and approving these plans (the Department of Health Services) had little sympathy for local autonomy and threatened to reject all 58 plans. As of spring 1993, the new Department of Toxic Substances Control Division had approved 20 plans and rejected 34 (10 of which were resubmitted). Four counties withdrew their plans. The Tanner process was widely discredited by respondents as being "a good idea, but impossible to implement." In short, Tanner's innovation was to strike a structural compromise between state preemption and local NIMBYism. It failed because Tanner planning was not in itself sufficient for breaking political gridlock without prior changes in local attitudes, levels of trust, and waste-producing consumption.

Since the early 1980s, the forestry case has also evolved, and new policy experiments are just now being proposed in the northern California natural resources arenas. Brokered settlements, bioregional management, "summits," and "grand accords" have all been attempted as ways to involve the antagonists, allow a greater role for rural counties, and decrease the angry polarization that characterizes forestry disputes in California and Oregon.

Bioregional management has emerged as the most visible forest management alternative, along with the Forest Service's New Perspectives program. A strong component of both is the notion that forest watersheds and other forest regions sharing common topographies and ecosystems should be managed in a coordinated manner different from those of the state, private, and national forest systems. There are as many different conceptions of bioregionalism as there are proponents and critics of the plans. One ongoing approach—loosely referred to as "bioregional resource management in the Klamath province" (California)—began as a memorandum of understanding among a consortium of California state resource agencies, federal land managers, and some of the larger environmentalist and industry players. Bioregionalism in the Klamath "province" has led to increased communication—especially on harvest planning—between the various land managers. Surprisingly, some of the opponents at these ad hoc meetings and "bioregional councils" have requested and obtained quasi-decision-making powers over several disputed harvest plans for sites in northern California counties.

But power may be difficult to wrest from public land managers. Several forest planners and supervisors at the federal and state levels have said that bioregional management offers the advantage of increasing coordination between different landholders, and that the point of bioregionalism is *not* that power or ownership will devolve to the bioregion but that planning at the national forest level should be done in concert with local private landholders and large private timber companies. Thus, timber harvest plans on adjacent public and private lands would take into account a neighbor's harvest plans and projects. Ken Slater, a forest planner in California's Klamath National Forest argued in favor of bioregional management because of the improvements in information he expects from it:

> What does it mean to the Forest Service? It increases the ability to coordinate management, across time and space, with different property owners. If we knew what other people are doing in a watershed, we could know if actions would be sustainable. . . . I don't see anything in bioregionalism eroding private rights or Forest Service discretion. . . . we would do a better job of scheduling activity across time and space. We would know their plans and they'd know ours.

Local officials and community residents have looked upon bioregionalism with a jaundiced eye, suspecting that it may be "another ploy" to reduce their influence on local land uses. Clearly, their conception of bioregionalism has more teeth in it, and in their worst fears county supervisors, loggers, and some environmentalists worry that bioregional councils would have discretion over forests and watersheds currently held by the state and counties.

But not all proposals for regionalism in forestry management are aimed at merging existing political boundaries. Mendocino County is attempting to dramatically increase its role in influencing local forest practices. The county can do so under the state Z'Berg-Nejedly Forest Practices Act of 1973, which

> grants individual counties the right to recommend additional rules and regulations for the content of timber harvesting plans and the conduct of timber operations to take account of local needs. If the [state] Board [of Forestry] finds the recommended rules and regulations are consistent with the intent of the Forest Practices Act and necessary to protect the needs and conditions of the county recommending them . . . such rules will be adopted within 180 days. (Burkhardt, 1992, p. 1)

The Mendocino County Board of Supervisors, however, narrowly turned down the recommendations of its Forest Advisory Committee calling for the county to limit harvest rates and practices and, in essence, to exceed the state's rules and regulations. Moreover, since the California Department of Forestry (CDF, which implements the Forest Practice Act) can veto county rules, local discretion may be limited.

Despite the possibility for county-level control, many respondents still feared that a bioregional council would take authority away from counties—in effect, that bioregionalism was just another way for the state to gang up on local communities. And rural communities already blame their ills on the larger state and federal political units, with their predominance of "proenvironmentalist urbanites." Like any other policy proposal in the forestry case, bioregionalism will have to squarely face the conundrum of the urban-rural split, collective versus individual interests, and competing ecological goals. More broadly, the success of any policy innovation depends on how well it addresses concerns over social justice, technocracy, and economic imperatives. In chapter 6 I return to the three challenges to democracy to see if they seemed to constrain policy-making in the two case studies.

CHAPTER SIX ❦ Challenges to Democratic Environmental Policy-Making in Forest and Hazardous Waste Management

How do the empirical data fit with the challenges to democracy described in chapter 3? Do these challenges help explain people's political choices and preferences, or do the respondents' answers offer refutations to these challenges? For example, a choice to use litigation and injunctions would probably not be described by many as a "highly participatory" strategy, but those who choose such strategies may see them as the surest means available to reach a socially just outcome. Others might refute the importance of the "temporal" aspects of the environmental problem for democracy (i.e., that democratic decisions take longer to reach) by pointing out that (1) conflict will continue regardless of the environmental urgency of certain issues (in fact, conflict will probably increase), and (2) policy-making by consensus, or some approximations thereof, requires less time for implementation than conflict-ridden policies. This chapter assesses the three challenges to democracy in light of the empirical research performed for this study.

The Social Justice Challenge

Since no Environmental Czar or Supreme Tribunal was appointed or created during the period in which these two cases were researched, does that mean that the social justice challenge was an empty threat to democratic policy-making? Not quite, because the social justice challenge could manifest itself in the form of preferences for environmentalist guardians rather than a full-blown environmental Leviathan. Recall that a social justice challenge could work through a judiciary that was simultaneously insulated from current democratic structures (to say nothing of participatory governance) and charged with protecting some form of "environmental justice." And, whatever its merits, a hierarchical judicial system is not participatory democracy. In fact, persuasive arguments have been made by rural sociologists (Carroll, 1992) that judicial and legal processes of environmental litigation and hearings are skewed toward elite lobbyists and professionals who can understand and manipulate such systems.

It would be simplistic to conclude that the use of *any* litigation at all in these two cases could demonstrate the strength and presence of the social justice challenge. Instead of tallying the number of times opponents go to court, it is more useful to look at the role the courts play in people's strategies—are they willing to "take a chance" on getting a sympathetic hearing in a court of law? A strong preference for litigation may indicate a social justice challenge, especially when more participatory forms of conflict resolution are considered a waste of time. This is clearly the case with Andy Kerr (of the Oregon Natural Resources Council), whose principal tool is litigation.

Kerr has much more faith in the courts than in mediation or hearings because he has had substantial victories there: "The proof that our information is good lies in the Dwyer decision. ONRC's information meets a high standard to prove that the Forest Service made arbitrary and capricious decisions and that it violated the law. . . . our lawsuits are not frivolous, they would lose if they were."

Here Kerr equates court successes with technical and legal validation, saying, in effect, "If I win, I must be right." By extension—given his pragmatism in the search for the most successful venues for achieving his environmental goals—Kerr is also saying, "If the court decides in my favor, it is the right arbiter of these issues." Of course, this is not the same as saying that the court should make new decisions, not based upon legislation, concerning the spotted owl or timber harvesting. Relying on the courts is not the same as calling for enlightened guardians to manage in the absence of democratic procedures; it only indicates an overriding preoccupation with environmental outcomes that presages support for, at least, quasi-guardianship models.

But Kerr was not typical of all the respondents or even of all the environmentalist respondents. Many of the latter saw social justice, empowerment, participation, and healthy environments as inextricably linked. Most of the environmentalist respondents, along with loggers in Hayfork, California and Springfield, Oregon perceived their dilemmas to be the direct result of a lack of power and fairness. By and large, they had never had the experience of winning a major court battle at the bench of a sympathetic judge; indeed, the notion that their cause might be championed by a leader or a noble institution hardly crossed their minds.

There are thus really two results here: first, the social justice challenge is strongest when people use the courts to restrict further debate or action. Those who use the courts as their primary strategy tend to be more concerned with specific environmental outcomes than with political processes that emphasize participation. Not surprisingly, a second and conflicting conclusion can be drawn from the responses of those who were unwilling to separate "right outcomes" from what they saw as the most fair politics—participatory politics. This latter group wants to achieve empowerment at least as much as, if not more than, it does to reach specific environmental outcomes. Just as environmental pragmatists argued for any political strategy that could aid their environmental goals, the toxics activists and the loggers who wanted to gain influence welcomed whatever venue would empower them: courts, protests, media attention, NIMBYism, recalls, and hearings. A guarded conclusion, based on the two cases studied here, is that the social justice challenge is most pronounced where environmental goals are ranked above participatory political strategies, and the opposite is true when respondents either identified environmental outcomes with participation or ranked participation above—or prior to—environmental goals.[1]

This conclusion might surprise or even offend decentralists like Orr (1992), who argue that the strands of environmentalism form (or ought naturally to form) a coherent ideology. In such an ideology, "good" environmentalists favor the empowerment of the poor, self-realization, and community-building activities in addition to sound environmental policy. Kassiola (1990) echoes Orr's assumptions about the unity of environmentalist political philosophy and ideology.[2] However, the themes of Kassiola's and Orr's ecological politics make more sense as prescription than as description. Both recognize that much more can be done to cultivate a taste for participatory democracy. Orr's educational program for developing ecological literacy in fact calls for a great deal of learning about politics and the policy process. The key to diminishing the importance of the social justice challenge may be to educate participants in environmental dilemmas about the interaction of environmentalism and political values. This is especially true for situations where participants engage in single-issue politics.

The Technocratic Challenge

Was the technocratic challenge significant in the two cases studied? As discussed in chapter 4, the technocratic challenge has three major attributes:

1. The ability to fill a gap of *technical competence and information*
2. The tendency for environmental legislation to require only *pro forma public participation* of technocracies
3. The creation of *inequality and negative instead of positive authority* under technocratic management

Chapter 5 showed that bureaucracies were not the only ones to provide information in many environmental conflicts; activists, academics, legislative staff, and industrialists all had the ability—and often the willingness—to become expert themselves. As long as the "general public" turns to experts of one kind or another, these examples suggest that information may still be part of the technical challenge, because grass-roots activists who become expert may also be called technocrats—even though they are outside of the bureaucracy. The point is that participants still either (1) turn to expert consultants or (2) become expert themselves, and thus are no longer part of the "unwashed masses" of citizens. It *is* fundamental that many citizen-participants have recognized that their own technical competence is by no means cause for exclusion.

That said, participants all expressed a desire for more—not less—information concerning their specific conflicts. And although they felt that their opponents' information was suspect, respondents nevertheless expressed a desire to have their views supported by "hard science." Thus, most respondents could see that information was an indefensible barrier to participation and mistrusted many sources of that information, but continued to require rigorous evidence to support their arguments. As Jasanoff (1991) argues, this preference for hard evidence only strengthens the role of technocracy:

> The phenomenon of protracted technical controversy, accompanied by continual shifts from one to another locus of uncertainty, supports the view that quantitative risk assessment is far less an independent decision technique than a surrogate for deeper political divergences that choose

(or sometimes are forced by law) to express themselves as disputes about evidence. Yet if risk analysis merely converts political arguments into technical ones, the cause of accountability is not likely to be well served. Because of their esoteric nature, such methodologies threaten to elude the traditional processes of democratic control. Scientific peer review provides a partial remedy against crass abuses of quantitative techniques, but this is a remedy that tends, if anything, to consolidate legitimation of public policies, not least because peer review in the regulatory environment can so easily be subverted by politics. (p. 44)

Fundamentally then, conflicts over technical information strengthen the technocratic challenge by raising the ante of what Jasanoff calls "acceptable evidence," thereby maintaining a large role for experts even in the face of declining scientific credibility. However, as long as scientific information is disputed, technical experts cannot claim that "correct" policies exist independently of individual analysts. When technical expertise and rationality are not disputed, policymakers may be unable to exert control over their expert advisors "because the techniques employed by the experts assume that the 'one best answer' will be a function of techniques—not the individual expert."[3]

While more people can participate, it is not clear from the respondents' answers that their participation was seen as meaningful. Most of the agency personnel and legislative staffs in the forestry and toxics cases were reluctant to venture estimates of how effective the public's participation actually was. They felt that the formal procedures their agencies had adopted were implemented in good faith but that final decisions should remain theirs.[4] In the forestry cases, timber industry spokesmen and environmentalists agreed that the Forest Service's public "scoping" hearings were a waste of time and money, and that most of the important decisions seemed to have been made before the hearings were held and were altered little by them. Forest Service respondents conceded that they started holding public consultations (especially in the 1970s under the Resource Planning Act—see chapter 5) without either the expertise to run them smoothly or a clear idea of how to use public input. That said, all of the Forest Service supervisors and planners interviewed underlined the personal rapport they had developed with members of their communities.

But if all respondents viewed participation as a waste of time, why would they continue to try influencing environmental outcomes? In fact, there were benefits to maintaining personal contacts with agency staff, commenting on all timber harvest plans in the hopes of significantly influencing the outcomes of some of them (whether participants belong to SHARE or the Sierra Club), and showing up in large numbers at meetings. Luke Cole, of the California Rural Legal Assistance (CRLA), pointed out that if activists alone attend a hearing, their participation tends to be viewed as a legalistic process of only moderate concern to policymakers. When two hundred community residents show up, it becomes "political" and thus grabs the attention of elected officials.

The gap between pro forma participation and effective participation is not all that wide, but it changes depending on one's definition of "effectiveness." Dahl (1989) defines "effective participation" thus:

> Throughout the process of making binding decisions, citizens ought to have an adequate opportunity, and an equal opportunity, for expressing their preferences as to the final outcome. They must have adequate and equal opportunities for placing questions on the agenda and for expressing reasons for endorsing one outcome rather than another. (p. 109)

Dahl's emphasis on agenda setting and outcomes is important; if he is correct, then participants should be unsatisfied if their activities result in anything less than policy change (or resolute adherence to the status quo, depending on the participant in question). For example, timber industry respondents in northern California and in Oregon regularly show legislators their managed forests and facilities like the biomass waste-to-energy facility in northern California. They meet with members of Congress, they show them their lands, and they try to allay their concerns about industrial management. Environmentalists do the same thing. Each side claims that their representatives leave in some way influenced by what they've seen. If the standard of effectiveness is to determine a final decision, then taking legislators on a tour will rank low. But respondents justified their participation on more than one criterion of success, and that may explain why they continued to participate despite continued losses or resistance from policymakers.

Not surprisingly, the question of effective participation is closely related to mobilization for negative versus positive authority. As Mazmanian and Morell (1990) put it:

> Having an extremely active citizenry at the local level seems like the epitome of democracy to most Americans. But when, in effect, they only have the power to stop proposed facilities, the process is far from complete. Few mechanisms have yet been devised to bring competing interests together in ways that allow them to identify their common goals, narrow their differences, and move forward to identify and site in acceptable locations those genuinely needed industrial facilities and public facilities that today are being vetoed. No viable democratic process exists to mesh the collective needs with those of individuals and localities. (p. 134)

Logically, if participants cannot achieve their goals working with agencies and technocrats, they may resort to stopgap actions, at least to prevent a deterioration in their state of affairs. Where participation is perceived to be ineffective, such stopgap measures are likely to be in opposition to current policies and projects because of the many administrative avenues for delay open to participants. In none of the cases did the respondents characterize their endeavors as "constructive" or as a way of moving progress forward as opposed to "holding the line." Respondents across the spectrum did talk about what would be more constructive: building natural resource or technological eduction in the hope of developing a more sympathetic general public, or restoring eroded forestlands and increasing salmon runs in wild rivers (see the discussion of "success" in chapter 5). But despite their visions of more positive outcomes to community empowerment, these were not the activities that occupied most of their time as political participants.

Assuming that negative authority was much more prevalent in respondents' strategies than positive authority, was that negative authority largely a function of technocracy or of general dissatisfaction with an unresponsive political process? In the case of toxics, neighborhood organizers mobilized to oppose externally imposed risks, not to build hazardous waste treatment solutions, because siting issues were forced upon them. And as Elliott (1984) points out, many opponents of treatment facilities are more concerned about safe management and emergency response than they are about the technology per se. They do not trust the technocrats who advocate facilities or the managers who will run them.

In the forestry case, negative authority was probably due to perceptions of political weakness as much as distrust of technocrats. In forestry disputes, conflicts look much more like traditional distributive struggles over scarce resources. If opponents cannot work through the political process to build positive, community authority but are able to block policymaking, they will do so. In the absence of leadership from legislators, heads of large environmental organizations, and corporate officials, respondents looked to agency officials or experts neither for solutions to political gridlock nor for ways to reverse the uses of negative authority. Thus, the respondents' only alternative to negative power was to eschew *any* political influence at all, and that was an unacceptable choice.

A final component of the technocratic challenge was technocracy's commitment to technology—not just any technology but higher levels of technical innovation fueled by social competition and inequality. Note that Leiss's (1972) and Borgmann's (1988) claims are reminiscent of Hirsch's "positional economy": a "status" economy and popular belief that economies must grow, quantitatively, to raise the living standards of the worst off support the pursuit of gains made in technological sophistication. Society seeks these gains not to increase the abilities of citizens to enlarge the scope and effectiveness of their participation but rather to keep them striving toward a standard of living that is materially impossible for them to achieve.

Was this aspect of the technocratic challenge present in these two cases? Inequality and technology played a greater role in the toxics case than in the forestry case, if only because hazardous waste really is a by-product of technology and industrial society. In the toxics case especially, activists argued for "appropriate technology," source reduction, and more education on the links between routine economic consumption, technology, and hazardous wastes. Most people familiar with hazardous waste issues agree that some major trade-offs in consumption patterns would have to be accepted in order to eliminate the production of all, or almost all, hazardous wastes. Luke Cole, CRLA lawyer, emphasized that Kettleman City residents may be more willing to site a toxics incinerator "when it's safe enough to site in Beverly Hills!"

Many toxics activists felt that although they had not shared in the prosperity created by industrial technology, they were being asked to shoulder industry's burdens nonetheless. Moreover, the army of experts involved in industry—and in industry's regulatory constraints—have not attempted to mitigate the ills of hazardous waste treatment problems by turning their expertise toward socially benign technologies or nontechnical means of diminishing inequality. Technocrats seem to operate on the assumption that most aspects of industrial society are fixed or are not worth attempting to change. Hence their commitment to a technical solution—hazardous waste incineration—whose social implications are as important as its technological ones.

The Economic Challenge

This study was conducted during a period of national economic recession that hit California particularly hard. The cancellation of defense contracts, increased high-tech competition from Japan, and logging injunctions handed down by federal courts were just some of the factors that consistently kept unemployment in California one to two percentage points higher than the national average. And with the state running the largest budget deficit in its history, it is not surprising that economics permeated many of the interviews conducted for this study, affecting the strategies for and conceptions of desirable—and feasible—environmental policy.

Despite the perennial finding that a majority of the public thinks much more should be spent on the environment (or *any* amount necessary, for that matter), the mood among respondents was strongly pro–economic growth. Even the environmentalists conceded that they could no longer term their demands as absolute necessities, *whatever the costs*. When asked about the viability of logging communities and timber companies, Jeff DeBonis, founder of the Association of Forest Service Employees for Environmental Ethics (AFSEEE), tempered his impatience to consider the all-important question of jobs and revenue: "I say roll 'em [the logging communities and timber companies]. But we don't *have* to roll 'em. In the short term, we'll have a downturn, but also a more labor- and human-intensive silvaculture. [We should] pay people for their labor, not corporations for their machinery." DeBonis went on to explain that his organization had given a lot of thought to economic questions, by formulating economic transition programs aimed at both loggers and timber industries. It would be unthinkable for DeBonis and the AFSEEE to ignore economic considerations in the policies they urge, even though they are convinced that the Forest Service and timber industries are lawbreakers.

In the forestry example, the economic challenge to democracy threatened to restrict the policy debate to only those policies that would essentially leave logging communities financially intact. Indeed, the Forest Advisory Committee of Mendocino County argued for adoption of more stringent rules largely because it projected sharp declines in employment, county income, harvest levels, and county harvest revenues if current forestry practices continued.

But how could one of the highest paid (and least educated) blue-collar workforces be protected? Environmentalists and industry struggled to answer that question in order to gain an upper hand in defining the range of possible policy options. Some, like Dennis Gilbert, who argued that a labor-environment coalition was possible, were trying to show that the economies of small communities need not be hurt by reducing timber harvesting:

> It is the common interest of the Environmental and Labor Movements to fight for alternatives for workers and communities so that the economic blackmail is reduced, and to otherwise fight for political and economic rights for working people so that they are not tied to the political outlooks of their employers. Or to put it another way, environmental reform puts the larger issue of economic transition on the table. Inasmuch as we fail to recognize and take responsibility for this, we don't have our feet on the ground of reality, much less the common ground of the Labor and Environmental Movements. And we will be divided and conquered by people pursuing their short term greed.[5]

Environmental policies that could strongly affect the financial health of many people were considered to be inappropriate intrusions by the loggers and timber industry respondents. As Hochschild (1981) points out, in American politics, fairness does not necessarily dictate that economic policies and outcomes be equally open to all those seeking influence. In the forestry case, this meant that the environmentalists had to legitimize those of their demands that would affect rural economies; as mentioned, some attempted to do so by mitigating their policies, but others downplayed the sanctity of jobs over all else. DeBonis argued that he could talk about "rolling" logging communities because the United States had experienced such shocks in the past: "We've *had* economic transitions across the U.S. . . . auto makers put 30,000 people out of work without recoil. . . . it's accepted under capitalism."

The timber industry response in this example from John Hampton, not surprisingly, was to discount the value of economic transitions and to argue that catastrophes would result from more restrictions on harvesting: "In Oregon we have more than 70 communities that depend exclusively on forest products for their livelihood . . . and those are, by and large, suburban communities where they don't have a choice. . . . retraining is an illusion."

The owner of the biomass energy facility in northern California painted a grim picture of the future for all natural resource–dependent communities:

> Where I think it leads us, in the near term, [is] to massive unemployment fairly quickly, potentially a depression, and then, probably a realization that we have gone too far. That we never want to return to the old days where we abuse the resources in the course of utilizing them, but certainly a middle ground where we say "hey, we know enough now to utilize these resources wisely and protect them in perpetuity." But the trouble with that system, that scenario, is that there's tremendous pain for hundreds of thousands, potentially millions of people that come out of that before we arrive back at the point where conservation, in its true definition, and the wise use of natural resources becomes, basically, the norm.

Another way for environmental activists to counter the economic challenge in the forestry case was to broaden the focus of their appeals for increased participation to include as wide a jurisdiction as possible, preferably, at a national level. There, they could count on auto workers in Michigan who would be fairly disinterested in a few thousand jobs out West and on a lot of urban support for "green cathedrals."

In the toxics case, the economics of hazardous waste management worked for both the industrialists' and the environmentalists' goals. There were three reasons for this:

1. The cost of facilities, or of community compensation, was much less of an issue than was siting approval.
2. For at least one of the industries, Dow Chemical in Contra Costa County, economic incentives for hazardous waste management pointed toward small-scale incineration *after* an aggressive source reduction program was implemented.
3. Together with NIMBYism, state and federal bans on land disposal of

hazardous wastes coincided to create mounting pressures on companies to reduce their wastes.

The economic challenge was not so present in the toxics case, mostly because environmentalists had succeeded in overshadowing economic questions (e.g., transportation, treatment, storage, and disposal costs) with health-based moral claims. This finding affirms Portney's (1984) claim that people are more concerned with safety than with financial remuneration—suggesting that, if anything, money should be put into source reduction, and increasing the sophistication and safety of facilities, rather than invested in community centers. But even a state-of-the-art facility will not ensure siting approval. Sophisticated hazardous waste incinerators have been rejected by local communities in Kettleman City, California, in Arizona, in Mexico, and Canada, and in many other parts of the United States (Rotella, 1992; Morell, 1992). When asked if design considerations might affect their decisions, respondents in Livermore and Kettleman City, California all claimed, "It shouldn't be here. Period."

But the comments of one Department of Toxic Substances Control (DTSC) official may foreshadow a future where resources are so scarce that hazardous waste management options will be sharply curtailed:

> I think we're entering a period of accelerating scarcity. It's not going to be a gradual process, especially in California. . . . Ten years from now California's going to be a dramatically different place if we continue with this business-as-usual response way of dealing with issues. And consequently, there's going to be a whole new awakening for activists in terms of risk prioritization. It's inevitable. . . . there's going to be a compaction, and the allocation of resources to marginal activities is going to disappear. And in toxics what we're doing right now is putting together a comparative risk study to really say "where should we be directing our resources?"

Site cleanups like the one in Sunnyvale proved to be more affected by costs than did incinerator siting dilemmas. Neighborhood residents like Bonnie Bradshaw did not secure their cleanup option, precisely because costs were deemed too great, she claimed. But EPA officials argued that excavation to eight feet was sufficient and that cost considerations did not enter into EPA's decision because no amount of excavation could physically remove all of the PCB laden sludge at the Westinghouse site without dispersing it further—potentially into the aquifer the agency was charged with protecting.

Of course, there is no way of knowing with certainty if the EPA ever really considered the 50-foot option, but economic considerations may play a greater role in site cleanups than in the siting of treatment facilities precisely because most residents want *some kind* of cleanup anyway. Thus, their acquiescence to an EPA cleanup project is already in part secured—in fact, the whole cleanup process is likely to have been initiated by local residents. Ironically, residents filed some of their numerous lawsuits against Westinghouse, arguing that the contamination at the Westinghouse site had depressed their property values; however, they had to drop their suits when several independent property assessments and a few sales demonstrated less financial harm than had been originally estimated (Russell, 1992).

The Dow Chemical plant located in the Pittsburg/Antioch area of Contra Costa County almost built a small hazardous waste incinerator on-site in 1992–93, but the project's demise shows how a participatory process always risks failure if a single key group "defects." Once again, economic considerations did not play a primary role in the politics of this incinerator project. Although the incinerator faced initially weak opposition and then acquiescence by statewide and national environmental representatives, local opposition coalesced against it at the eleventh hour. Dow's posture was quite conciliatory as the company organized its strategy for obtaining community and agency approval: First, Dow Chemical was very careful to circumscribe the scope of the project to treating its own on-site wastes. Dow was able to point out that if only on-site wastes were treated, transportation risks associated with sending wastes to Louisiana would be reduced. Second, Dow worked with the local Greenpeace chapter to implement a source reduction program that dramatically cut hazardous waste production at the site. Dow was thereby able to demonstrate that it had not oversized its incinerator; in fact, the company could show that source reduction would remain a first option at the facility because every pound of waste burned incurs a treatment cost and a cost in chemical product lost to the waste stream. As Dow Chemical's Fischback put it, "With an offsite commercial incinerator, in general, the more waste produced by another generator which you burn, the more money you make. This would provide a strong incentive to increase capacity." Since Dow was proposing to build an *on-site* facility, the company could avoid this tendency to increase capacity. Finally, Dow claimed that if local residents had strongly resisted siting the facility or had demanded that an even smaller incinerator be built, Dow would have continued to ship the wastes out of state. The company was not prepared for a protracted struggle and thus came to the bargaining table with a slightly more flexible agenda. In contrast, spokespersons for Chem Waste Management in Kettleman City affirmed that *not* siting their incinerator was "non-negotiable."

There's no question that people's fears of health risks have dampened industry's claim that new treatment facilities were an economic imperative for industry's survival. However, a potentially positive result of NIMBYism and land disposal bans is that economics have constrained *industry*'s policy options in ways that may be beneficial. Provided treatment costs continue to go up and enforcement of waste tracking improves, California's per capita waste stream may continue to drop before new treatment capacity is installed in the state.

Several legislative staff and industrialists argued that not enough attention is paid to the economic costs of environmental policies. Catherine Morrison, chief of staff to Republican Assemblywoman Cathy Wright (Simi Valley), felt that environmentalists had grown powerful enough to ignore business costs when drafting toxics legislation:

> In this legislature the environmental community is not required to back up their positions with detailed policy analyses. . . . when you have too much power on your side, you don't have to justify what you're doing, [and] that makes for bad policy. I know when I've been challenged to justify our positions . . . when I sat down with the bill sponsors, [the bill's justification] wasn't there . . . [but] Republicans have felt that environmental regulations have been burdening business for no good reasons, they [often] don't make the environment better.

In these two environmental controversies the bottom line is still very much an economic one. The economic challenge to participatory environmental decision making is probably the most difficult challenge of the three discussed in this study. Consensus and compromise is so difficult to reach in environmental conflicts like the two presented here because people perceive real power not in terms of participation and citizenship but in terms of their economic future. Industries in both cases were unwilling to share power not because there was something inherently wrong about participation but because they felt that most of the policy options advanced by their adversaries would make their prospects for economic growth uncertain. Economic issues seem more tangible than political ones; indeed, the political realm is often perceived to be "unfair" in its distribution of goods and opportunities, in contrast with the market that is considered "fair and wise" (Lane, 1986, p. 385). Indeed it may be that democracy and environmentalism are compatible but that American liberal democracy and *economic growth* are destined for endless conflict.

Just as truly participatory experiments are rarely attempted in American politics, innovations in economically attractive environmental policies are only beginning to be implemented. There is no shortage of policy proposals that rationalize economic concerns with environmental ones, but environmentalists are rarely willing to impute "environmental values" to industry, and industry rarely trusts that environmentalists can truly take economic concerns to heart. At the level of rhetoric, then, the economic challenge succeeds in restraining policy debate by characterizing traditional opponents as *incapable* (as well as unwilling) of truly considering alternative policy options. The economic challenge, at its strongest, is thus symptomatic of a lack of imagination, trust, and empathy.

CHAPTER SEVEN Conclusion

Comparing Survey Results to In-Depth Interviews

The results of the in-depth interviews discussed in chapters 5 and 6 did not differ greatly from general trends reported in environmental opinion research. Most respondents in this study echoed the support for participation that Pierce, Lovrich, and Matsuoka (1990) found in elites, activists, and the general public. What the in-depth interviews also revealed was that people will support participation conceptually as long as its meaning remains abstract. But when they were presented with various implications of increasing participation, support among many respondents diminished. The in-depth interviews with respondents in the toxics and forestry cases showed that people were more concerned with their own participation than with someone else's. In fact, questions that asked about people's "support for participation" were almost banal, as if the term "participation" was just a generic and universally accepted good. Who could make friends arguing against participation? Public opinion research is inherently limited in its ability to link multiple thoughts and comments from respondents, and this prevents students of environmentalism from following up on the meaning of "support for participation." In-depth interviews showed that most people were dissatisfied with the results of their participation. Only a few large surveys have shown that people tend to support participatory strategies less than adversarial ones (e.g., litigation) if by doing so they may lose some decision-making power or influence over their particular conflict.[1]

Conceivably, if the industrialists, legislators, and agency officials interviewed in this study had filled out a questionnaire like the one used by Pierce et al. (1990, see chapter 4), they might indeed have been counted as the elites who overwhelmingly supported more participation on environmental policies. As in the Pierce study, when asked "generically" if they supported "more local participation" on their issues, they responded affirmatively, with only a few exceptions. But further discussion revealed that they were uncomfortable with extending anything more than a limited advisory role to the general public. Even if they conceded that the public had influenced them to significantly alter management decisions, they were reluctant to formalize and transform public influence into quasi-permanent structures empowered with specific oversight or vetoes. The one exception was a memorandum of understanding (MOU) signed by state and federal resource agencies with jurisdiction for the "Klamath province." The MOU commits these agencies to solicit and use an unprecedented degree of public input. And if large timber companies continue their exodus to Chile, British Columbia, and the southeastern United States, resource managers may be freer to allow local decision making.

The role of information in environmental disputes has also been largely ignored by survey researchers, perhaps because they hope that focusing on scientists and the public's trust in experts will capture the essence of technical information and its impact on participation (i.e., less trust in scientists might lead to more involvement, and vice versa). An exception was the study by Pierce (1989) examining the level of policy-relevant knowledge on environmental issues held by people in Shizuoka, Japan and Spokane, Washington. Pierce's respondents were fairly well informed and did not experience technical complexity as a barrier to participation, and these findings were supported here.

In general, the results of this study expand upon, rather than refute, those of survey research. In some areas, like the relationship between economic growth and environmental protection, the comfortable margins of support for spending vastly more on the environment, as reported by survey researchers, seem to erode in the face of actual policy decisions. Future survey research should incorporate questions designed to further measure people's *personal* willingness to consider alternative policies in light of explicit economic considerations (e.g., scenarios of environmental costs). Research should then follow up to see how many people would act on alternative policies, given the opportunity.

Sustainable Policies or Participatory Democracy?

The beginning of chapter 3 described two evaluative criteria by which to judge the two environmental cases in this study. The first prescribes examining the policies developed and implemented in both cases to see if they meet some measure of "sustainability." The second involves showing how these policies were developed, through means that were relatively participatory or not.

Not surprisingly, the cases studied here did not meet the stringent criteria outlined in chapter 3. There were no great breakthroughs in consensus that allowed for the kind of long-term planning needed to sustain diverse natural resources and healthy (or healthful) industries. In toxics management, the best that can be said about California's efforts over the last ten years is that industries in the state have reduced their waste streams by approximately 5 percent (Thurm, 1991). But the state of California still cannot account for at least 19 percent of the wastes generated in state (Schmitt, Carey, and Thurm, 1991a); only one in four sites slated for cleanup by the state has been certified as clean (Schmitt, 1992); and maybe half of the state's "recycled" wastes are burned as fuel oil with little or no air quality controls (Schmitt, Carey, and Thurm, 1991b). According to Kip Lipper (1992),[2] treatment capacity for new wastes has not increased enough to handle existing wastes or the new wastes that are still being produced. Meanwhile, researchers are still in sharp disagreement over the dangers of hazardous wastes.

One could argue that, in the absence of direct correlations between groundwater contamination and health effects, hazardous waste landfills in remote, dry areas ought to remain open. But this would not meet the requirement that management efforts seek to avoid irreversible effects. In fact, support for the status quo in hazardous waste management means that landfills and certain groundwater basins would be condemned in perpetuity.

Superfund sites in California have fared no better, and most of the time, much worse. It is likely that a Superfund site will not be cleaned up for at least ten years after it was listed (Thurm, 1992; Plater, Abrams, and Goldfarb, 1992). However, hundreds of "emergency sites" are minimally cleaned up—or excavated—on very short notice each year (McKinley, 1992). Of the 28 Superfund sites in Silicon Valley, soil has been removed from 18, extraction wells have been drilled for 25, and half have been assigned treatment programs (U.S. EPA, 1989).

Recent environmental policy implementation in the forestry case is also hard to characterize as sustainable. Policies have mostly lurched from one piece of legislation to a court decision, and back again, as opponents square off to save or log one more forest. Increasingly, the Forest Service and other agencies are being led to adopt a more preservationist rhetoric, but changes in allowable sale quantities (ASQs) and other major policies still appear to be driven by conflict litigation. Left to itself, the Forest

Service would probably change its practices slowly, especially as a different White House administration replaced the Service's top policymakers and Congress mandated lower timber sales. The various advocacy coalitions and policymakers involved in forest management have not been able to reach or implement strategies that forgo permanent changes to the most unique forest ecosystems. Meanwhile, policies that would save the way of life of rural communities have not been seriously considered by the agencies—or Congress—under fire from both pro- and antilogging forces.

Ecological restoration might provide high-paying outdoor jobs to loggers and millworkers, but the Forest Service and other resource agencies are not free to consider these kinds of policies, even if they agree with them. The same problem confronted the personnel of the Army Corps of Engineers in securing nonstructural alternatives to dam construction in the 1970s (Mazmanian and Nienaber, 1979). Restoration work could help to restore the diversity and longevity of riparian and forest habitats, as well as help repair the erosion damage from the largest clearcuts. Restoration and selective, low-impact logging could provide enough jobs for the relatively small forest products industry workforce, but they would have to be paid for by higher timber prices, greater mechanization, more efficient management practices, or wage subsidies of some kind.

This is not to say that policy experiments have not been tried in either the forestry or toxics cases. Experiments designed to promote both sustainability and participation have been attempted, especially in California, over the last ten years (see chapters 5 and 6). As discussed earlier, these include bioregional management, Tanner planning, county and regional hazardous waste management commissions, partnerships, compacts, and the Sierra/Grand Forest Accords of 1991–92.

Each has had an important impact on the course of political activities within the forestry and toxics arenas. But as policy experiments, they have rarely been successfully implemented, or even implemented at all. Although Tanner plans were submitted by all of California's counties, the same disagreements and gridlock that occurred before Tanner planning still exist between the state and local governments and the residents. The Sierra Accords, bringing industry and environmentalists together to work out timber harvest reforms on private lands, failed when they were plugged into partisan and regional legislative politics.

And for every policy experiment there are opposing forces that seek to reduce public input. For example, the Bush administration sought to reduce public comment on timber harvest plans, arguing that "there is ample opportunity for public input under all our environmental statutes. . . . But in some instances we feel there are duplications that have led to long delays. The fact is that in administrative appeals there is an opportunity for giving any citizen the right to tie up projects for an interminable period" (Schneider, 1992). Congressional delegates from the Pacific Northwest also routinely attempt to limit judicial review of timber harvest plans in riders to appropriations bills (Durbin, 1990c; Plater et al., 1992).

Many policymakers think participation and sustainability are inextricably linked, as the authors of the global report *Our Common Future*[3] (1987) point out over and over again. By the logic of the tragedy of the commons, all those who affect their local environments must be involved in protecting them.

The report, issued by the World Commission on Environment and Development (WCED), assumed that participation is at least a necessary, though not sufficient, condition of sustainability. Viewed in this light, the efforts of participants in this study's two cases should not be judged too harshly. It is true that if one looks for comprehensive change toward sustainability, the record appears grim, but on an incremental level, many small efforts are continually being made to pump and treat groundwater supplies, to limit the size of clearcuts, to recycle hazardous wastes and wood products, and to keep updating the concept of sustainability. The findings in this study give no indication, however, that the overall trend in resource and pollution management leans more toward than away from sustainability.

Conclusions about the Centralists and Decentralists

There are compelling elements to both centralist and decentralist analyses; it is easy to understand why each conveyed a strong message. In their simplest expressions, both offer convincing accounts of human nature. The centralists emphasize the urge to survive, the natural need for order and security, and the attractiveness of leadership (especially in times of crisis). The decentralists stress people's equally strong desire for self-determination and their conviction that humans, like any creatures, are loathe to foul their own nests.

But how useful are the centralists' and decentralists' positions for understanding specific environmental issues, such as the ones that arise in the cases of this study? Are they right or wrong, or does this debate confuse more than clarify? Three themes emerge in response to this question: (1) in their most typical forms, both the centralist and decentralist analyses are too thinly developed to work as political prescriptions; (2) people often react to environmental crisis by increasing their participation, not necessarily by demanding more centralized control; and (3) people mobilize more to prevent losses than to make gains.[4]

Political Prescriptions

First, the blanket formulae suggested by the centralists and decentralists are indeed problematic and difficult to apply to many environmental cases. Local and decentralized approaches can result in widely varying standards as different localities weigh environmental priorities in different ways. Typically, communities that acknowledge their own ability to bear the economic and political burdens of environmental protection, and that have the means to do so, will do more to prevent health risks and ecosystem degradation. Poorer communities (or rich communities less sympathetic to environmental concerns) might fear the costs of proenvironment policies (see chapter 3). Thus, the decentralized prescription, in its simplest expression, relies heavily on a fortuitous congruence of enlightenment (or at least informed discourse) and economic well-being.

The centralist approach to environmental policy-making really does risk compromising fairness and can inhibit participation. Large organizations have had only mixed success managing environmental problems, usually succeeding only when such problems were relatively discrete in scope. The most visible examples are reductions in the use of lead paint and leaded gasoline, CFCs in aerosol cans, DDT pesticides, and PCBs in transformers (Commoner, 1987). But environmental problems are almost always interconnected—with other resource and pollution problems or with economic and social issues. Complex problems that involve numerous disparate interests and high stakes are not often successfully addressed by *any* management scheme, much less by centralized resource and pollution agencies (Dryzek, 1987a).

A hybrid approach tends to achieve the most common denominator (the status quo) and encounters huge coordination problems between local and centralized power. For example, the two cases in this study would not present identical opportunities for management by a centralized-local hybrid, but what criteria should be used to decide on a meaningful and effective mix of local and central control? If "local knowledge" (e.g., oral history) is crucial to any resource management, including centralized agencies, how should it be coordinated with expert research (e.g., remote sensing)? Moreover, local knowledge is silent on many issues; in these instances, which source of knowledge should prevail?

Even in a centralist-decentralist hybrid, many of the people affected by environmental policies will judge the process by its outcome. If it appears that the centralist partner (e.g., the Forest Service) in the hybrid structure makes the final decision, observers may conclude that the decentralized unit (e.g., bioregional council) actually has no influence in that particular environmental policy arena. Of course, that conclusion may not be completely accurate, especially if, as often happens, the decentralized unit had dissuaded the administering agency from implementing key provisions of an initial plan. Agency officials interviewed for this study often pointed out that they had altered their plans or scrapped projects entirely based on public consultation. Officials typically do not perceive that they have very much discretion or power and thus may consider changes in their decisions to be substantial, even if they appear small or inconsequential to outsiders. In any case, it is very hard to prove that agency officials changed their policy directions, a problem noted by Mazmanian and Nienaber (1979):

> To what degree did the district [Kansas City District Office of the Corps of Engineers] intend to share its powers? . . . The KCD made it clear that it would not be bound in any formal way by the decisions of the study groups or wishes of any of the citizen groups. But the fact that the study groups were convened and that the public was continually encouraged to contribute to the planning process surely implied, at least to the participants, that they would share in making the decision . . . We can only

surmise from the outcome—the district's tentative compromise proposal and then inaction—that citizen input was indeed taken seriously and that some power was shared. (p. 100)

As mentioned earlier, perhaps the centralist-decentralist dichotomy confuses more than it clarifies; effective participation on environmental decisions should be measured along a spectrum between these two poles, and the degree of local versus centralized control should be evaluated very much in the context of specific cases.

Sometimes the shortcomings of centralized, decentralized, and hybrid systems may be mitigated. For example, an important attribute of centralized power is its ability to offset some of the opponents' losses in environmental disputes. The federal government can sponsor economic transition programs, offer more incentives for American timber milling, or even purchase entire towns condemned by hazardous wastes.

Crisis Participation

The second theme that emerges from observation of the centralist-decentralist dynamic is that the centralists seemed to miss the possibility that environmental crisis or urgency would motivate people to participate—not to call for some environmental czar. Although the environmental effects of clearcutting and toxic spills may be profound, trees and toxics controversies are not "loaded" with the kind of urgency that characterizes a famine, flood, or the ozone hole. Thus, the panic and perception of crisis that Ophuls (1977) and Heilbroner (1980) believed would drive authoritarian politics has not materialized in these environmental issues; on the contrary, grass-roots participants involved in these issues are *increasing* their participation and resisting various forms of co-optation, or centralized rule.

In fact, as the interviews showed, people tried to address environmental issues through the existing structures of liberal democracy and found the system wanting. People did not immediately chain themselves to trees to get results; they first put pressure on Congress and state legislators to correct environmental problems by state and federal law. It is only *after* the failure of such major attempts at reform through the institutions of representative democracy that there was a sharp rise in NIMBYism and civil disobedience or protest organized (pro or con) around environmental issues.

Part of the centralists' analytical problem stems from using an external threat, like war, as a metaphor for environmental predicaments that are really generated within the society. For example, Heilbroner predicted that environmental crises would increase as people experienced growing civilizational malaise, apathy, and resignation to a long period of decline. People have already responded to perceived crises to a greater degree than he predicted, but they still lack the ability to coordinate their actions.

This difficulty in coordinating participation into effective action, coupled with the perceived failures of national legislation, accounts for some of the NIMBY syndrome but also shows how participants can conclude that both political leadership and the political process have failed them. Leadership has failed because national legislation did not translate strong public support into lasting gains, and the political process has failed because it has become overtaxed and gridlocked instead of responsive.

In sum, the centralists took a holistic look at a set of problems that had to be addressed and were overwhelmed. But on a local level, many people perceive these problems as smaller and more discrete, some of them even manageable. Moreover, participants could distinguish between process and outcomes: public involvement occurred with specific projects (harvest plans or particular toxics treatment facilities) and with the *process* of decision making used by agencies. For example, John Teie, a forester for the CDF, said that his agency can take these process comments at any point but that project-specific questions are harder to integrate, especially if they come at a late stage in a harvest plan decision. Reviews of timber harvest plans are usually on a very tight schedule that starts when a plan is filed, and then runs at 10 to 20-day intervals between each consultation step in the process. The whole decision process can be over in a one-month period—not because the agency is rushing a decision but because it is mandated by state law to act within a specified period of time. His observations suggest that project-specific comments could be better handled by the agency if timetables had more flexibility, thereby allowing participants to comment on all aspects of timber harvest planning.

Loss Aversion and Mobilization

A third theme is that people mobilized against losses rather than for some anticipated gains. As was discussed in chapters 5 and 6, they were motivated to participate because they thought they were losing important material, political, economic, or social goods. Such activism is marked by a sense of dread and crisis rather than a positive sense of community building or even of general progress.

This theme echoes recent research on risk perception and risk aversion. Quattrone and Tversky (1988) have shown that, for most of the respondents they surveyed, potential losses were generally perceived to be more salient than potential gains: "An important consequence of loss aversion is a preference for the status quo over alternatives with the same expected value" (p. 724). They studied people's risk aversion when both losses and gains were at stake and found that most respondents wanted higher probabilities of success when negative outcomes were at stake than they did for potential windfalls. This was true despite the fact that the same probabilities existed for both losses and gains (suggesting that perhaps faulty personal calculations could also explain some of the risk aversion). Most important for this study, Quattrone and Tversky stressed the role of loss aversion in conflict resolution:

> Loss aversion may play an important role in bargaining and negotiation. The process of making compromises and concessions may be hindered by loss aversion, because each party may view its own concessions as losses that loom larger than the concessions of the adversary. . . . This difficulty is further compounded by the fact . . . that the very willingness of one side to make a particular concession . . . immediately reduces the perceived value of this concession. (p. 726)

And if environmental issues are more often viewed as struggles against losses than activism for gains, then the stakes in environmental politics may appear greater than they might be for other political issues, where interests vie for more traditional, *distributional* gains. Thus, the effective pursuit of negative power (e.g., NIMBY) may be a special characteristic of environmental problems that makes them so controversial. To the extent that this negative power could be turned into constructive mobilization, it might defuse some of the controversies enough to break the political gridlock over issues similar to both of the cases in this study. Part of this conversion of the environmental dilemma may be achieved by turning participation into a consensus-building exercise with real gains rather than an ongoing effort at damage control.

Converting negative authority into positive authority may be possible in the area of environmental regulatory enforcement by moving to "positive action compliance and clean-up" (Mazmanian and Morell, 1992). Positive action compliance offers opportunities for a regulated community to perceive and obtain real benefits from compliance, for example, by creating win-win situations that may save targeted businesses money while they comply with regulations.

The Future of Participatory Environmental Policy-making

The analysis of the preceding chapters lends itself to exploring prescriptions for encouraging participatory environmental policy-making and to determining where these might be possible and appropriate. A key question is, Are there governance structures or policy issue characteristics that lend themselves to participatory environmental policy-making? Several theorists have tried to answer parts of that question by looking at the nature of political conflicts in specific settings; these were discussed in chapters 2 and 3, and throughout this study.[5]

A common feature of many of these analyses of policy change and institutional design is their focus on information, authority, and problem characteristics. So much political and social criticism faults contemporary governance systems and the citizens that live in them for their poor information-processing capabilities and remote, elite-based decision making tendencies. Theorists like Hult and Walcott, Warren, Sabatier, and Ostrom are not arguing that the solution to these systemic problems is merely to turn up the volume on participation but rather to draw attention to the social bases of authority. Instead of uniformly prescribing decentralized participation, they ask what features of problems and issues lend themselves to participatory problem solving.

This was the thrust of Warren's emphasis on social goods discussed in chapter 2. For Warren (1992), there are two necessary elements to expansive democracy: the *agents* of political change and the *social problems* they attempt to address. First, citizens must be able to transform them-

Table 7.1 Design Characteristics of Participatory Environmental Policy-making

DESIGN ISSUES	CRITERIA	DIMENSIONS
Characteristics of Environmental Problems or Goods	Do they lend themselves to incremental, decentralized management?	Acute, chronic, complex, bounded, reversable, excludable or nonexcludable goods, solubility
Policy and Process Learning	1. Do participants and affected citizens have policy-relevant knowledge? 2. Are they familiar with the policy process? 3. Is policy-relevant information mistrusted; is there learning among advocacy coalitions?	Limited level of detail reached on issue-knowledge or policy process; sophisticated understanding, but entrenched beliefs; negotiable knowledge and information (e.g., value-change is possible)
Jurisdictional Rules and the Agents of Policy Change	1. Who should be included as participants in decisions? 2. What institutions should be the foci of policy action?	Varies depending on how participants are affected by issues, how well participation can be coordinated across physical and political boundaries, legal requirements
Ecological Rationality	Are the policy options either under consideration or available for implementation environmentally sustainable?	Policy reversability, information-processing capacities, resilience, extent of problem displacement

selves into effective political actors. Second, not all problems will serve as vehicles for self-transformation. The benefits of some social goods and policy issues are not especially divisible, and thus participation merely aggravates the struggle over scarce resources.

As it applies to specific environmental issues, Warren's analysis is appropriately incomplete due to its focus on the general characteristics of self-transformation and democracy. But a new account of the "best" institutional and policy design strategies for environmental cases will probably remain very contextual (i.e., of low generality) until issue characteristics, information, and authority are all taken together in institutional prescriptions. This means that innovative policy and institutional designs will have to consider (1) how specific issues constrain policy choices, (2) how participants and nonparticipants perceive, and react to, technical information (and information about the policy process), and (3) how jurisdictions and rules of inclusion and exclusion can be agreed upon so that the agents of policy change can be identified with a reasonable degree of consensus. Table 7.1 illustrates key questions that should be asked of participatory environmental policy-making designs.

The first column lists the design issues that successful participatory environmental policies should address. The second column shows what evaluative criteria can be used to determine if design issues are reflected in governance structures. The third column describes the spectrum of dimensions along which design characteristics probably vary. Evaluating governance structures—and designing new ones—in light of the characteristics listed in Table 7.1 remains an essential research agenda for studying participation in environmental problems.

AFTERWORD The Distant Democracy

I' th' commonwealth I would by contraries
Execute all things; for no kind of traffic
Would I admit; no name of magistrate;
Letters should not be known; riches, poverty,
And use of service, none; contract, succession,
Bourn, bound of land, tilth, vineyard, none;
No use of metal, corn, or wine, or oil;
No occupation; all men idle, all;
And women too, but innocent and pure;
No sovereignty.
—William Shakespeare, *The Tempest*, act 2, scene 1

Gonzalo's utopia was unsullied by the evil practices and governments of human society; it was "natural" and derived its goodness from nature's—not man's—imprint. But Gonzalo's natural society could only be conceived in a societyless island, washed up on the edges of a New World. Shakespeare thought he chose well when he chose the New World as a place free from the fetters of society and want, but America was never empty, and its nature was never so benevolent.

American political culture has always made us think we were somewhere else, or that we were on our way to a place different from where we had been before. Somehow the American landscape would do for people what our society and the Old World could not. But our democracy remains distant. Dreams of a land of enlightened citizens and meaningful participation are constantly disappointed, even though American politics is steeped in the rhetoric of participatory democracy: community, town meetings, accountability, self-determination, the voice of the people.[6] Indeed, why do so many voters stay home, only to clamor for their say when they are threatened by some crisis? Why have so many people come to distrust the hearing room, the advisory committee, or the local resource agency? Perhaps it is because their participatory ideals are constantly left unfulfilled: the seat on the board really does not change much, public advisory structures result in policy changes only after what seems like years of struggle, and so on.

This country has never truly tested participatory democracy in environmental dilemmas. Effective participation, including power-sharing and decision-making authority, has not actually been attempted in either of the cases in this study. No treatment facility sponsor ever agreed to share management responsibilities. In the forestry case, the story is the same: Congress has never seriously considered devolving discretion over national forests to locally interested parties.

It is surprising that this simple fact has escaped the scrutiny of political theorists. At first the observation seems naïve. After all, we know what would happen if neighborhood activists had a say in deciding siting questions, don't we? We assume that, because of NIMBYism, they would never site a facility, and we are sure that locals are incapable of taking some responsibility for hazardous wastes.

It is certainly hard for individuals and local organizations or governments to see a fair way for them to accept a role in managing hazardous wastes; the problem seems so enormous, so intractable, so caught up in the remote industrial machine of the American economy. And residents of poor communities can reasonably wonder why all "consumers" must bear equal responsibility for wastes if their own consumption patterns are so different from those of wealthier communities.

Perhaps consumers, rich and poor, might feel motivated to accept responsibility if they had the opportunity to shape more aspects of modern society—for example, by squarely addressing consumerism instead of the palliatives that merely redirect its appetite. As it is, their choices appear to be stark and few: a toxics incinerator *or* dangerous waste stockpiles, a pristine forest *or* a tree farm. These are not perceived as choices but as impositions. Real choices would allow people to comment on more fundamental, higher order questions, like Do we want to consume products that create hazardous wastes? or Can we build with less wood and less waste so that fewer trees need to be cut from our public lands? Participatory environmental policy-making can be written off as an ineffective way to make decisions only if participants experiment—and fail—with these questions.

But participatory democracy's highest value is just generic empowerment. To successfully save both democracy and environmental value, democratic participants must learn to know and accept that they live in partnership with each other and with nature. Living together requires social justice; living with nature requires environmental sustainability.

And living apart from ourselves or nature is not a life society can nurture for very long. If environmental management goes the way of the Leviathan, it will not be for the sake of environmental prerogatives but as an excuse to impose order, as Leiss's analysis (1972) suggests.

After three decades of environmentalism, it appears that not just ecological things are all tied together, so are politics. The distant democracy is hard to reach because it takes individual *and* collective action to get there, individual value changes *and* social learning. The choice is not really between environment and democracy; hopefully, the failures of the first three environmental decades will make that clear for citizens of the 1990s and beyond.

Notes

CHAPTER ONE Environmentalism Returns to the Democratic Fold

1. See Hardin, 1968; Heilbroner, 1980; Ophuls, 1977; and Ophuls and Boyan, 1992.

2. See Taylor, 1992; Eckersley, 1992; Orr and Hill, 1978; Sale, 1991; and Schumacher, 1973. Of course, the centralist and decentralist positions occupy two endpoints of a spectrum, and are useful insofar as they explain the *tendencies* of policy advocates who fall somewhere in between.

3. AB 2948; Chapter 1504, Statutes of 1986; Health and Welfare Code Section 25135 et seq. and other sections.

4. Now plans must be sent to the Department of Toxic Substances Control (DTSC), the office in the new Cal-EPA responsible for managing hazardous waste rules and regulations.

5. See the typologies of environmental worldviews in Lester and Dryzek, 1989.

CHAPTER TWO Environmental Political Thought and Democratic Theory

1. As Taylor paraphrases the options, "Is political democracy of greater value than appropriate environmental policy or vice versa?" (1992, p. 2).

2. Walker, 1988.

3. Taylor, 1992.

4. Lester and Dryzek, 1989.

5. Heilbroner, 1980, p. 128.

6. Heilbroner, 1980, p. 175.

7. Heilbroner, 1980, p. 172.

8. See my discussion of the "economic challenge to democracy" in chapter 3 of this book.

9. Ophuls, 1977, pp. 185–86.

10. A steady-state economy would exist, as well as possible, within the physical limits of the earth and would recycle waste with minimum loss. Waste outputs would be recycled into physical systems with high efficiency, while the extraction of nonrenewables would be sharply curtailed. See Daly, 1980, for further details.

11. Ophuls, 1977, p. 163.

12. Passmore, 1974, p. 183.

13. Passmore, 1974, p. 183.

14. Orr and Hill, 1978, p. 463.

15. Orr and Hill, 1978, p. 466.

16. "Bioregionalists" have long argued that decentralization is at the core of the decentralist analysis and prescription. In their view, a sustainable environment and society can *only* be achieved in small, semiautonomous regions delineated by features of the "natural" environment (topography, watersheds, ecosystems). For a good introduction to the normative aspects of bioregionalism, see Sale, 1991.

17. Orr, 1992, p. 31.

18. Rodman, 1980, p. 72.

19. Rodman, 1980, p. 74.

20. Orr, 1992, p. 34.

21. Rodman, 1980, p. 68. Nazarea-Sandoval, 1994, elaborates on this theme by showing how farmers in the Philippines deliberately plant a wide range of sweet potato varieties that are no longer in favor with government extension services. By hiding their old varieties between rows of the government's hybrids, these farmers engage in "everyday forms of resistance" that are as effective as they are undramatic.

22. Passmore, 1974, p. 195.

23. Eckersley, 1992, p. 20.

24. Sabatier, 1988, points out that policy networks or subsystems "should be broadened from traditional notions of 'iron triangles'—limited to administrative agencies, legislative committees, and interest groups at a single level of government—to include actors at various levels of government active in policy formation and implementation, as well as journalists, researchers, and policy analysts who play important roles in the generation, dissemination, and evaluation of policy ideas" (p. 131). I would add to the advocacy coalition concept *any* local activist, commission member, business leader, or other participant who follows policy developments and consistently tries to influence them.

25. Sabatier, 1988, p. 155.

26. For example, the appeal of California's Proposition 65 (1986) was that citizens would strike a bold, comprehensive blow against exposure to toxics without suffering dilution from compromise in the legislature. Another big environmentalist success with the initiative process was passage of Proposition 70 in 1988. Proposition 70 was a $776-million bond initiative issued to preserve wildlife habitat and to acquire parks and recreation lands. Similarly, environmentalists attempted sweeping changes in Proposition 128 (nicknamed "Big Green"), the 1990 proposition that would have reformed pesticide use and a plethora of other environmental policies.

27. Caves, 1992, found that over 90 percent of the citizens he interviewed agreed that "public meetings should be called to discuss the merits and drawbacks of a proposed initiative once it gets the necessary signatures" (p. 22). Caves thought that his respondents perceived this proposal as a "forum to eliminate

frivolous initiatives and also an opportunity to get some first-hand information on the issue being addressed by a specific initiative" (p. 23).

28. Many of today's state legislatures could, however, be faulted for the same gridlock and dearth of reasoned deliberation.

29. Though Gross, 1980, argues that most electoral participation is just a smokescreen for the real powers, mostly corporate and technocratic, that form "the Establishment." Thus, any electoral process can be abused by elites and power brokers:

> The more that people are encouraged to "throw the rascals out," the more their attention is diverted from other rascals that are not up for election: the leaders of macrobusiness, the ultra-rich, and the industrial-military-police-communications-health-welfare complex. Protests channeled completely into electoral processes tend to be narrowed down, filtered, sterilized, and simplified so that they challenge neither empire nor oligarchy. (p. 240)

30. See Fishkin, 1991, chapter 8, "New Structures of Representation: Deliberative Opinion Polls and Other Proposals."

31. Burnheim, 1985, has suggested that the best way to reconcile the scale of polities with policy complexity is to do away with electoral systems altogether. He proposes functional representation instead, whereby citizens' names are randomly drawn to serve on any number of limited-function commissions or councils. While their names would be chosen at random, these citizens would have to meet certain criteria to show that they were sufficiently affected by the scope of the commission they would join. Thus geographical location, age, and occupation, among others, might ensure that potential members serve because they are suitably representative of the citizens affected by the corresponding commission.

One major problem with Burnheim's proposal is that it really assumes that most issues can be dealt with on a local basis. Furthermore, he does not offer a way of coordinating functional councils across jurisdictions, much less across different issue-areas. Problems such as ozone depletion could well be impossible to address with functional councils, yet would strongly affect everyone under the jurisdiction of such bodies.

32. Expansive democracy is both transformative and highly participatory.

33. Warren, 1992, p. 11.

34. Warren, 1992, p. 12.

35. Warren, 1992, p. 12.

36. Her cases ranged from groundwater management in southern California to overfishing off the Turkish coast, wood removals in small woodlots of rural Japan, and grazing rights in alpine Switzerland.

CHAPTER THREE The Challenges to Democratic Environmental Policy

1. See Dryzek, 1987a, for the evaluative criteria by which he judges "ecological rationality." Any ecologically rational social choice mechanism, he argues, must be capable of (1) negative feedback, (2) coordination, (3) flexibility, (4) robustness, and (5) resilience.

2. The human component of this definition of environmentalism avoids the strongly bio- or ecocentric strands of environmental thought and activism (Eckersley, 1992; Scarce, 1990; Fox, 1990; Devall and Sessions, 1985) not because these critics have little to offer, but rather because their demands are certain to be too stringent for virtually the entire American political spectrum. Using deep ecology as an evaluative criterion would guarantee that the politics of the two environmental case studies would fail to be called "environmentally successful."

3. See Sale, 1980, pp. 301–6; Daly, 1980; Schumacher, 1973, pp. 40–52; Ophuls, 1977, pp. 184–98; Heilbroner, 1980, p. 98.

4. See Sale, 1980, pp. 156–64; Schumacher, 1973, pp. 146–59; Winner, 1986, pp. 22–39; Ophuls, 1977, pp. 156–64; Orr and Hill, 1978, pp. 461–63; Fischer, 1990, p. 44.

5. See Wandesforde-Smith, 1990, pp. 325–30, 339–43; Schumacher, 1973, pp. 53–62; Heilbroner, 1980, pp. 128–35, 177; Orr and Hill, 1978, p. 464; Eckersley, chapter 1.

6. The answer will depend partly on which assumptions about human nature one holds and on what problem-solving possibilities different forms of governance might have.

7. Ophuls, on the other hand, squarely rejects such an authoritarian world, particularly because he believes the centralized option to be incapable of truly achieving a steady state.

8. Dahl, 1989, points out that quasi-guardians are unlikely to protect minorities for very long. Consider the case of the Supreme Court:

> The quasi guardians of the Supreme Court rarely hold out more than a few years at most against major policies sought by a lawmaking majority. What the American experience indicates, then, is that in a democratic country, employing quasi guardians to protect fundamental rights from invasion by the national legislature (as distinct from state, provincial, cantonal, or municipal legislatures) does not provide a promising alternative to democratic processes, except in the short run. (p. 190)

9. Justice William O. Douglas actually published a book entitled *A Wilderness Bill of Rights* while he was on the Supreme Court. In it he reviews the mounting evi-

dence of pollution and legislative efforts made to curb it up to 1965, but falls short of enumerating a set of constitutional rights associated with the environment.

10. There are strong reasons for strengthening the judiciary's role in environmental decision making. As Sax foresaw in 1970, judges typically stand "outside" of political processes that bias politicians' decisions, and they are not likely to spend a great deal of their time on environmental controversies. Thus, a judge's positions on environmental matters is unlikely to sink his or her appointment. The judicial process also "demands that controversies be reduced to concrete and specific issues rather than be allowed to float around in the generality that so often accompanies public dispute" (Sax, 1970, pp. 104–12).

11. See Sax, 1970, chapters 4 and 11; Plater, Abrams, and Goldfarb, 1992, chapter 11, section D, "Citizen enforcement and judicial review."

12. This process is reminiscent of Dryzek's, 1990 communicative rationality.

13. But Laski, 1931, maintains a central role for the expert:

> What can be done is not what the expert thinks ought to be done. What can be done is what the plain man's scheme of values permits him to consider as just. . . . the more closely the public is related to the work of expertise, the more likely is that work to be successful. For the relation of proximity of itself produces conviction. The public learns confidence, on the one hand, and the expert learns proportion on the other. Confidence in government is the secret of stability, and a sense of proportion in the expert is the safeguard against bureaucracy. (p. 13)

14. To extend the analogy, it may suffice for the doctor to adopt a "good bedside manner" in order to win the trust and confidence of the patient; failing that, the patient may want to seek a second opinion from a very different kind of practitioner.

15. Indeed, *whether* the state may legitimately impose coercion as a result of rational discourse is open to question.

16. Activists might argue that this is because their counterparts are supposed to be representing them in the legislature.

17. The feller-buncher is a good example—this machine can cut and remove trees in one function, with far less labor. It is also a complex piece of machinery, costing up to several hundred thousand dollars.

18. Freeman, 1989, suggests that, of the various forms of democratic capitalism, those with corporatist-mixed political economies seem to be most democratic because they can ensure real representation and clout as permanent bargaining members. They will be especially so if their electoral systems are more consensual than majoritarian because a greater number of views will be represented in coalition governments. He argues that corporatist-mixed economies will also be able to offer the greatest intragenerational and intergenerational wealth to the labor force,

because corporate owners will be less likely (and able) to pick up and move (collective gains being made societywide rather than regionally), and representatives can work out longer-term wage benefits because of their "staying power."

19. As for centralist solutions, the economic challenge also demonstrates that the dominant paradigm of growth economics constrains not just democratic decision making but the policy choices of any political system. This is another reason that authoritarian systems are unlikely to be better equipped for environmental protection: they too are prisoners of the market.

20. *Seattle Audubon Society v. John L. Evans (U.S. Forest Service) and Washington Contract Loggers Association*, 1991, U.S. District Court, Western District of Washington, 771 F. Supp. 1081.

CHAPTER FOUR Empirical Study of the Centralist-Decentralist Debate

1. A more complete description of these attributes follows (Pierce, Lovrich, and Matsuoka, 1990):

1. Background attributes (education, age, gender, and income)
2. Policy-relevant knowledge
   —Level of self-assessed knowledge
   —Familiarity with technical terms
   —Knowledge of local environmental conditions
   —Command of environmental science facts
   —Awareness of policy area problems
3. Political orientation
   —Support for the New Environmental Paradigm (developed by Dunlap and Van Liere, 1978)
   —Postmaterial value orientations (based on Inglehart's method [1971] for determining the degree of preoccupation with material acquisition)
   —Support for science and technology
   —Preservationist identification
4. Policy-specific issues (especially local)
   —Relative seriousness of local environmental issues
   —Perceived danger to the environment
   —Satisfaction with local environmental policy
   —Perception of the overall quality of the local environment

2. A paradigm shift described by Catton and Dunlap that posits four assumptions about human nature, social causation, and the context and constraints of society. Humans in the NEP are just one among many interdependent species, human affairs are linked to nature in ways that may produce unintended conse-

quences to human action, society is embedded in a finite biosphere, and human inventiveness cannot transcend ecological "laws" and limits (see Buttel, 1987).

3. The question on participation was worded in such a way that respondents could be thinking about any number of structures, institutions, and political processes. The exact wording was:

> In recent years there has been considerable debate over the value of efforts to increase the amount of CITIZEN PARTICIPATION in government policy making in the environmental policy area. How would you locate yourself on the following scale regarding these efforts;

These efforts are of *no value* and add needlessly to the cost of government	—1 2 3 4 5 6 7—	These efforts are of *great value* even if they add to the cost of government

4. As part of his study, Cotgrove, 1982, probed his respondents on questions of economics and environmental policy. On a scale similar to the one used by Pierce, Lovrich, and Matsuoka, 1990, (seven points, strongly oppose to strongly support), he found the following distribution of either support (five points) or strong support (six to seven points) among his respondents for raising taxes for pollution control:

93.5% of environmentalists
88% of nature conservationists
63% of industrialists
78% of public officials
76.3% of trade officials
63.5% of the general public

When it comes to "jobs versus the environment," the picture changes a bit in favor of jobs. The numbers scoring five or greater in favor of protecting the environment are:

75.3% of environmentalists
68.2% of nature conservationists
40.5% of industrialists
43.3% of public officials
26.7% of trade officials
41.8% of the general public

And when asked to rank the relative importance of welfare, law and order, economy, energy, environment, and foreign affairs as "objectives for government action," environment ranks fifth or sixth for all but the environmentalists (who rank it third). The economy ranks second for environmentalists but first for all the others.

5. One-third to two-thirds of respondents in national polls agreed that economic growth could be sacrificed for environmental protection; see Dunlap, 1989.

6. A limited amount of additional survey data on trees and toxics is available for the state of California. San Francisco's Field Institute publishes the *California Opinion Index* (COI), a monthly survey of attitudes held by Californians on various policy issues. The most recent polls focusing on the environment were conducted in October 1984 (published in the January 1985 issue) and July–August 1986 (published in the September 1986 issue). The 1984 poll found that the three environmental issues ranked most important by Californians were (1) "insuring the purity of the state's drinking water" (86%), (2) "cleaning up existing toxic waste sites" (84%), and (3) "limiting the disposal of toxic wastes" (83%). In late 1984, forests fared differently: 48 percent of the respondents ranked "limiting timber cutting on governmental lands and wilderness areas" as an issue of "high importance," 43 percent gave it "moderate importance," and 7 percent said it had "low importance." In this poll, there were more respondents who rated the state's performance on toxics negatively than who were concerned about forestry issues. Seventy-five percent of the respondents said the state was doing a fair (26 percent) or poor/very poor (49 percent) job of limiting the disposal of toxic wastes; only 14 percent said the state was doing a poor/very poor job of limiting timber cutting on governmental lands and wilderness areas. Unfortunately, we know little about which groups in California held these views. The COI asked, "Do you consider yourself an environmentalist?" and divided the answers according to four state areas: the San Francisco Bay, Other Northern California, Los Angeles/Orange, Other Southern California. If the category "Other Northern California" had been limited to rural northern California and if answers had been cross-referenced to the respondent's location, we might know if the mixed support for limiting timber harvests occurred because of residents in the northern counties.

In the 1986 poll, toxic waste was the single issue out of 26 statewide concerns that respondents ranked most important (74 percent said they were "extremely concerned" about toxic wastes). "Protecting the state's environment" was ranked eighth, with 63 percent of the respondents extremely concerned. There may be some bias introduced into this survey, however, because the respondents were asked to rank their concerns from a prepared list; they did not have a chance to volunteer policy issues. Any number of issues might have displaced toxics, or other environmental issues (like forestry) could have been heavily weighted.

In a recent study of citizen opposition to a salt dome hazardous waste treatment facility in Dayton, Texas, Wright, 1991, found that most people surveyed were intensely suspicious of government and industry but still considered scientists fairly credible. Moreover, credible scientists had to be "independent—not company employees" in order to be trusted by residents. The federal government didn't fare much better. Only 32 percent of those surveyed were willing to agree that "federal government legislation in recent years has dramatically improved the safety and effectiveness of hazardous waste management methods and practices" (p. 57). Fifty-two percent of the sample disagreed and 16 percent didn't know. As for the salt dome technology proposed for the waste storage site, two-thirds of the respondents found it an unsafe and ineffective method of disposal. In summary, trust proved to play a more important role than concern about technical competence in the Dayton study.

Kraft and Clary, 1991, studied citizen participation and NIMBYism on siting of a nuclear waste repository. They wanted to test the conventional view of NIMBYism, which they characterized as a function of (1) distrust of project sponsors, (2) limited information about siting issues, (3) attitudes toward the project that are local and parochial and that ignore broader ramifications, (4) an emotional orientation toward the conflict, and (5) a high level of concern about project risks (pp. 302–3).

To test the validity of this assessment of NIMBYism, the authors analyzed the testimony of 1,045 individuals who participated in Department of Energy (DOE) hearings in Wisconsin, Maine, North Carolina, and Georgia. The DOE hearings were held in 1986 to receive input from communities on the siting of a nuclear waste repository. For present purposes, it is not necessary to go into the details of the authors' methodology; suffice to say that their content analysis revealed that "those who testified were moderately knowledgeable about technical problems of waste disposal, and they held broader geographic orientations and were less emotional than the model predicts" (p. 318). Participants did express a high level of risk aversion and distrust of DOE, but these opinions were based more on past practices and a moderate level of technical knowledge than on fears and emotions uninformed by facts.

Perhaps the most interesting finding, given the questions about local control that follow in this study, was that most participants expressed concern for a large geographic area, not just their own city or surroundings. Sixty percent of those making statements were concerned about their state as a whole, another 28 percent were worried about other states and the nation, and 11 percent mentioned international difficulties of nuclear waste management. Only 23 percent spoke solely of local impacts.

7. The complete results were as follows:

MOST IMPORTANT USE OF THE FOREST	1989	1990
Source of jobs and revenues	42%	39%
Source of paper and wood products	19%	18%
Home for fish and wildlife	13%	16%

MOST IMPORTANT USE OF THE FOREST	1989	1990
Recreation area	10%	12%
Source of fresh water	6%	6%
Scenic attraction	5%	6%
None	5%	4%

Note: Sample size was 600 respondents, with a margin of error of 4%. *Source* Durbin, 1990b.

8. Complete figures were:

	PERCENT AGREE		PERCENT DISAGREE		PERCENT UNSURE	
Cutting Trees . . .	1989	1990	1989	1990	1989	1990
Causes serious damage to fish and wildlife	50	51	49	44	1	5
Is a major source of water pollution	25	26	66	66	9	8
Is necessary to keep forests healthy and productive	76	74	23	23	1	3
Should not be allowed near recreation areas	69	66	29	32	2	2
Is a major source of jobs and tax revenues	88	87	11	11	1	1

Source: Durbin, 1990b.

9. To the question, "How important is it to incorporate the views of each of the following in reaching a viable system of hazardous waste management?" they obtained the following responses:

	BUSINESS	ENVIRON-MENTALISTS	HEALTH ORGANIZATIONS
Percent of respondents answering "very important"	92	82	80

	STATE OFFICIALS	LOCAL OFFICIALS	ORDINARY CITIZENS
Percent of respondents answering "very important"	72	86	61

Source: Mazmanian, Stanley-Jones, and Green, 1988.

10. To answer these questions Sabatier, 1988, recommends that studies of advocacy coalitions be conducted over a period of a decade or more; such longitudinal studies of specific policy arenas could help establish the link between citizen learning and policy outcomes.

11. See Hochschild, 1981, pp. 17–26.

CHAPTER FIVE Trees and Toxics

1. See Morone, 1991.
2. See Press, forthcoming.
3. The chemical industry responded to this assertion by arguing that small facilities handling only small amounts of wastes are unlikely to be economically viable.
4. Special assistant to state senator Barry Keene, one of the major legislative players in northern California forest management disputes.
5. See Dunlap's reviews (1991a and 1991b) of public opinion polls.
6. Associate environmental consultant, retired December 1991.
7. Although other key factors were probably more responsible; see Mazmanian, 1991.
8. See Mazmanian, Stanley-Jones, and Green, 1988; Lester and O'M. Bowman, 1983; Piasecki and Davis, 1987; Davis and Lester, 1989; Mazmanian and Stanley-Jones, 1991.
9. Mazmanian, Stanley-Jones, and Green, 1988.

CHAPTER SIX Challenges to Democratic Environmental Policy-Making in Forest and Hazardous Waste Management

1. Estrich, 1993, argues that when courts are expected to dispense social justice instead of individual justice, elected representatives can take inflammatory posi-

tions without having to be held accountable. Whatever decision a court reaches on a controversial matter, politicians can deny their own responsibility for public policy, especially when the public is primed to interpret a message broader than the confines of individual cases at hand. When policy advocates limit their strategies to the courts, they may lose the ability to influence political actors outside of the judiciary.

2. See the table comparing the politics of industrialism and the politics of ecology in Kassiola, 1990, p. 205.

3. Jenkins-Smith, 1990, p. 46.

4. And indeed, unless formal institutional jurisdictions are changed, it is their legal responsibility to make these decisions.

5. Transcript of a presentation to the Lane County Chapter of the Association of Forest Service Employees for Environmental Ethics (AFSEEE) on May 23, 1990.

CHAPTER SEVEN Conclusion

1. See Mazmanian and Nienaber, 1979, for a discussion of public participation in projects of the U.S. Army Corps of Engineers. Mazmanian and Nienaber frequently make the point that people's evaluations of their participation often depended on whether they felt they had influenced Corps decisions ("shared power") and whether they had attained the level of participation that they thought was appropriate and desirable for themselves.

2. Assistant to California assemblyman Byron Sher, 21st District.

3. In their 1987 report, the World Commission on Environment and Development (WCED) listed seven goals of sustainable development:

> 1. A political system that secures effective citizen participation in decision making
> 2. An economic system that is able to generate surpluses and technical knowledge on a self-reliant and sustained basis
> 3. A social system that provides for solutions for the tensions arising from disharmonious development
> 4. A production system that respects the obligation to preserve the ecological base for development
> 5. A technological system that can search continuously for new solutions
> 6. An international system that fosters sustainable patterns of trade and finance
> 7. An administrative system that is flexible and has the capacity for self-correction (p. 65)

4. This point is well supported by the literature in psychology on risk avoidance (see Quattrone and Tversky, 1988).

5. See especially the discussions of Dryzek, 1987a and 1987b; Sabatier, 1988; Warren, 1992; Ostrom, 1990; and Mazmanian and Nienaber, 1979.

6. See Morone, 1991.

Bibliography

Ahmad, Y. J.; El Serafy, S.; and Lutz, E., eds. 1989. *Environmental Accounting for Sustainable Development*. Washington, DC: World Bank.

Amy, D. J. 1987. *The Politics of Environmental Mediation*. New York: Columbia University Press.

———. 1990. "Decision Techniques for Environmental Policy: A Critique." In R. Paehlke and D. Torgerson, eds., *Managing Leviathan: Environmental Politics and the Administrative State*. London: Bellhaven Press.

Babbie, E. 1986. *The Practice of Social Research*. Fourth ed. Belmont, CA: Wadsworth.

Barber, B. 1984. *Strong Democracy: Participatory Politics for a New Age*. Berkeley: University of California Press.

Borgmann, A. 1988. "Technology and Democracy." In N. J. Vig and M. Kraft, eds., *Technology and Politics*. Durham, NC: Duke University Press.

Bramwell, A. 1989. *Ecology in the Twentieth Century: A History*. New Haven: Yale University Press.

Burkhardt, H. 1992. "Why Mendocino County Needs Additional Forest Practice Rules." Unpublished manuscript, January.

Burnheim, J. 1985. *Is Democracy Possible?* Cambridge, UK: Polity Press.

Buttel, F. H. 1987. "New Directions in Environmental Sociology." *Annual Review of Sociology* 13:465–88.

Carroll, M. 1992. Personal communication, February.

Caves, R. 1992. *Land Use Planning: The Ballot Box Revolution*. Sage Library of Social Research 187. Newbury Park, CA: Sage.

Cobb, J., and Daly, H. 1989. *For the Common Good*. Boston: Beacon Press.

Commoner, B. 1971. *The Closing Circle*. New York: Knopf.

———. 1987. "The Environment," *New Yorker*, June 15.

Cotgrove, S. 1982. *Catastrophe or Cornucopia: The Environment, Politics, and the Future*. Chichester, UK: Wiley.

Cronin, T. E. 1989. *Direct Democracy: The Politics of the Initiative, Referendum, and Recall*. Cambridge, MA: Harvard University Press.

Crouch, C. 1983. "Market Failure: Fred Hirsch and the Case for Social Democracy." In A. Ellis and K. Kumar, eds., *Dilemmas of Liberal Democracies*. London: Tavistock Press.

Dahl, R. 1971. *Polyarchy: Participation and Opposition*. New Haven: Yale University Press.

———. 1985. *Controlling Nuclear Weapons: Democracy versus Guardianship*. Syracuse: Syracuse University Press.

———. 1989. *Democracy and Its Critics*. New Haven: Yale University Press.

———. 1990. *After the Revolution?* New Haven: Yale University Press.

Daly, H., ed. 1980. *Economics, Ecology, Ethics: Essays toward a Steady-state Economy.* San Francisco: W. H. Freeman.

Davis, C. E., and Lester, J. P. 1989. "Federalism and Environmental Policy." In J. P. Lester, ed., *Environmental Politics and Policy: Theories and Evidence.* Durham, NC: Duke University Press.

deHaven-Smith, L. 1988. *Philosophical Critiques of Policy Analysis: Lindblom, Habermas, and the Great Society.* Gainesville, FL: University of Florida Press.

Devall, B., and Sessions, G. 1985. *Deep Ecology.* Layton, UT: Peregrine Smith.

Douglas, W. O. 1965. *A Wilderness Bill of Rights.* Boston: Little, Brown.

Dryzek, J. S. 1987a. *Rational Ecology: Environment and Political Economy.* Oxford and New York: Basil Blackwell.

———. 1987b. "Discursive Designs: Critical Theory and Political Institutions." *American Journal of Political Science* 31 (August): 656–79.

———. 1990a. *Discursive Democracy: Politics, Policy, and Political Science.* Cambridge, UK: Cambridge University Press.

———. 1990b. "The Environmental Politics of the Good Society." Paper presented for the workshop on ecology, Committee on the Political Economy of the Good Society, meeting of the American Political Science Association, San Francisco, August 30 to September 2.

Dunlap, R. E. 1989. "Public Opinion and Environmental Policy." In J. P. Lester, ed., *Environmental Politics and Policy: Theories and Evidence.* Durham, NC: Duke University Press.

———. 1991a. "Public Opinion in the 1980s: Clear Consensus, Ambiguous Commitment." *Environment* 33 (October): 10–15, 32–37.

———. 1991b. "Trends in Public Opinion toward Environmental Issues: 1965–1990." *Society and Natural Resources* 4: 285–312.

Dunlap, R. E., and Scarce, R. 1991. "The Polls—Poll Trends: Environmental Problems and Protection." *Public Opinion Quarterly* 55: 713–34.

Dunlap, R. E., and Van Liere, K. 1978. "The 'New Environmental Paradigm': A Proposed Measuring Instrument and Preliminary Results." *Journal of Environmental Education* 9:10–19.

Durbin, K. 1990a. "Oregonians Deeply Split over Owl." *Sunday Oregonian,* May 6.

———. 1990b. "Rangers Scramble to Meet Timber Quotas." *Oregonian,* October 15.

———. 1990c. "Politics Helped Delay NW Timber Management Plans." *Oregonian,* October 15.

Dworkin, R. 1985. *A Matter of Principle.* Cambridge, MA: Harvard University Press.

Eckersley, R. 1992. *Environmentalism and Political Theory: Toward an Ecocentric Approach.* Albany, NY: SUNY Press.

Elliott, M. P. 1984. "Improving Community Acceptance of Hazardous Waste Facilities through Alternative Systems for Mitigating and Managing Risk." *Hazardous Waste* 1 (3): 397–410.

Ely, J. H. 1980. *Democracy and Distrust.* Cambridge, MA: Harvard University Press.

Estrich, S. 1993. "Courting Danger." *Los Angeles Times,* March 21.

Field Institute. 1985. "Environmental Issues." In *California Opinion Index,* vol. 1, January.

———. 1986. "Public Concern About Important State Issues." In *California Opinion Index,* vol. 5, September.

Fischer, F. 1990. *Technocracy and the Politics of Expertise.* Newbury Park, CA: Sage.

Fishkin, J. S. 1991. *Democracy and Deliberation: New Directions for Democratic Reform.* New Haven: Yale University Press.

Fortmann, L.; Everett, Y.; and Wollenberg, E. 1986. "An Analysis of Timber Harvesting Plan Protests in California, 1977–1985." Unpublished manuscript, Department of Forestry and Resource Management, University of California, Berkeley.

Fox, W. 1990. *Toward a Transpersonal Ecology.* Boston: Shambhala Press.

Freeman, J. 1989. *Democracy and Markets: The Politics of Mixed Economies.* Ithaca, NY: Cornell University Press.

Gallup Poll. 1989. "The Environment." In *The Gallup Report,* no. 285, June.

Gormley, W. T. 1989. *Taming the Bureaucracy: Muscles, Prayers, and Other Strategies.* Princeton: Princeton University Press.

Gormley, W. T., and Peters, B. G. 1987. "Policy Problems and their Remedies." Paper presented at the annual meeting of the Midwest Political Science Association, Chicago, April 9.

Gross, B. 1980. *Friendly Fascism: The New Face of Power in America.* New York: M. Evans.

Hardin, G. 1968. "The Tragedy of the Commons." *Science* 162 (Dec. 13): 1243–48.

Heilbroner, R. 1980. *An Inquiry into the Human Prospect: Updated for the 1980s.* New York: Norton.

Hirsch, F. 1976. *Social Limits to Growth.* Cambridge, MA: Harvard University Press.

Hochschild, J. 1981. *What's Fair?: American Beliefs About Distributive Justice.* Cambridge, MA: Harvard University Press.

Hoffert, R. W. 1986. "The Scarcity of Politics: Ophuls and Western Political Thought." *Environmental Ethics* 8 (1): 5–32.

Hueting, R. 1980. *New Scarcity and Economic Growth.* Amsterdam: Elsevier North-Holland.

Hult, K., and Walcott, C. 1990. *Governing Public Organizations: Politics, Structures, and Institutional Design.* Pacific Grove, CA: Brooks/Cole.

Inglehart, R. 1971. "The Silent Revolution in Europe: Intergenerational Change in Postindustrial Societies. *American Political Science Review* 65: 991–1007.

Jasanoff, S. 1991. "Acceptable Evidence in a Pluralistic Society." In Deborah G. Mayo and Rachelled D. Hollander, eds., *Acceptable Evidence: Science and Values in Risk Management*. New York: Oxford University Press.

Jenkins-Smith, H. C. 1990. *Democratic Politics and Policy Analysis*. Pacific Grove, CA: Brooks/Cole.

Kassiola, J. J. 1990. *The Death of Industrial Civilization: The Limits to Economic Growth and the Repoliticization of Advanced Industrial Society*. Albany, NY: SUNY Press.

Kraft, M., and Clary, B. 1991. "Citizen Participation and the NIMBY Syndrome: Public Response to Radioactive Waste Disposal." *Western Political Quarterly* 44 (2): 299–328.

Lane, R. 1986. "Market Justice, Political Justice." *American Political Science Review* 80 (2): 383–402.

Laski, H. 1931. *The Limitations of the Expert*. Fabian Tract no. 235. London, UK: The Fabian Society.

Lauber, V. 1986. "Ecology Politics and Liberal Democracy," *Government and Opposition* 13: 199–217.

Leeson, S. M. 1979. "Philosophical Implications of the Ecological Crisis: the Authoritarian Challenge to Liberalism." *Polity* 11: 303–18.

Leiss, W. 1972. *The Domination of Nature*. New York: George Braziller.

Lester, J. P., and Dryzek, J. S. 1989. "Alternative Views of the Environmental Problematique." In James P. Lester, ed., *Environmental Politics and Policy: Theories and Evidence*. Durham, NC: Duke University Press.

Lester, J. P., and Bowman, A. O'M., eds. 1983. *The Politics of Hazardous Waste Management*. Durham, NC: Duke University Press.

Lindblom, C. 1977. *Politics and Markets*. New York: Basic Books.

———. 1982. "The Market as Prison." *Journal of Politics* 44: 324–36.

Lipper, K. 1992. Personal communication, March 12.

Loomis, J. 1987. "An Economic Valuation of Public Trust Resources of Mono Lake." University of California, Davis, *Institute of Ecology Report* no. 30, March.

Lowe, P., and Rudig, W. 1987. "Review Article: Political Ecology and the Social Sciences—The State of the Art." *British Journal of Political Science* 16: 513–50.

Luke, T. W., and White, S. K. 1985. "Critical Theory, the Informational Revolution, and an Ecological Modernity." In John Forester, ed., *Critical Theory and Public Life*. Cambridge, MA: MIT Press.

Lyden, F. J.; Twight, B. W.; and Tuchmann, E. T. 1990. "Citizen Participation in Long-Range Planning: The RPA Experience." *Natural Resources Journal* 30 (Winter): 123–38.

McCarthy, C.; Sabatier, P.; and Loomis, J. 1991. "Attitudinal Change in the Forest Service: 1960–1990." Paper presented at the annual meeting of the Western Political Science Association, Seattle, March 21–23.

McKinley, H. 1992. EPA staff, personal communication, April 23.

Magleby, D. 1984. *Direct Legislation: Voting on Ballot Propositions in the United States.* Baltimore: Johns Hopkins University Press.

Majone, G. 1988. "Policy Analysis and Public Deliberation." In Robert B. Reich, ed., *The Power of Public Ideas.* Cambridge, MA: Ballinger.

Mansbridge, J. 1980. *Beyond Adversary Democracy.* New York: Basic Books.

Maser, C. 1988. *The Redesigned Forest.* San Pedro, CA: R & E Miles.

Mazmanian, D. A. 1989. "California Toxics Policy and the Imperatives of Democracy." Paper presented at the annual meeting of the Northeast Political Science Association, Philadelphia, November 9–11.

———. 1991. "California Green: Toward the New Energy Paradigm." The Claremont Graduate School, *Center for Politics and Policy Report.*

Mazmanian, D. A., and Morell, D. 1990. "The 'NIMBY' Syndrome: Facility Siting and the Failure of Democratic Discourse." In N. J. Vig and M. Kraft, eds., *Environmental Policy in the 1990s.* Washington, DC: CQ Press.

———. 1992. *Beyond Superfailure: America's Toxics Policy for the 1990s.* Boulder, CO: Westview Press.

Mazmanian, D. A., and Nienaber, J. 1979. *Can Organizations Change? Environmental Protection, Citizen Participation, and the Corps of Engineers.* Washington, DC: The Brookings Institution.

Mazmanian, D. A., and Stanley-Jones, M. 1991. "Reconceiving LULUs: Changing the Nature and Scope of Locally Unwanted Land Uses." In Joseph DiMento and LeRoy Graymer, eds., *Confronting Regional Challenges: Approaches to LULUs, Growth, and Other Vexing Governance Problems.* Cambridge, MA: Lincoln Institute of Land Policy.

Mazmanian, D. A.; Stanley-Jones, M.; and Green, M. 1988. *Breaking Political Gridlock: California's Experiment in Public-Private Cooperation for Hazardous Waste Policy.* Claremont, CA: California Institute of Public Affairs.

Meadows, D. H.; Meadows, D. L.; Randers, J.; and Behrens III, W. W. 1972. *The Limits to Growth.* New York: Signet.

Morell, D. 1992. Personal communication, June.

Morone, J. A. 1991. *The Democratic Wish: Popular Participation and the Limits of American Government.* New York: Basic Books.

Muir, J. 1970. *The Treasures of the Yosemite.* Ashland, OR: Lewis Osborne.

Nazarea-Sandoval, V. 1994. *Lenses and Latitudes: On the Boundaries and Elasticity of Agricultural Decision Making.* Ithaca, NY: Cornell University Press.

Nelkin, D. 1977. *Technological Decisions and Democracy.* Beverly Hills, CA: Sage.

Oelschlaeger, M. 1991. *The Idea of Wilderness: From Prehistory to the Age of Ecology.* New Haven: Yale University Press.

Ophuls, W. 1977. *Ecology and the Politics of Scarcity.* San Francisco: W. H. Freeman.

Ophuls, W., and Boyan, A. S., Jr. 1992. *Ecology and the Politics of Scarcity Revisited*. New York: W. H. Freeman.

Oregon Business. 1991. "So Far Down It All Looks Up." August.

Ornstein, R., and Ehrlich, P. 1989. *New World, New Mind: Moving toward Conscious Evolution*. New York: Doubleday.

Orr, D. W. 1992. *Ecological Literacy: Education and the Transition to a Postmodern World*. Albany, NY: SUNY Press.

Orr, D. W., and Hill, S. 1978. "Leviathan, the Open Society, and the Crisis of Ecology." *Western Political Quarterly* 31 (Dec.): 457–69.

Ostrom, E. 1990. *Governing the Commons: The Evolution of Institutions for Collective Action*. Cambridge, UK: Cambridge University Press.

O'Toole, R. 1988. *Reforming the Forest Service*. Covelo, CA: Island Press.

Paehlke, R. 1990. "Lost Keys and No Engine: Re-Starting History in the Age of Ecology." Paper presented for the workshop on ecology, Committee on the Political Economy of the Good Society, meeting of the American Political Science Association, San Francisco, August 30 to September 2.

Paehlke, R., and Torgerson, D. 1990. "Environmental Administration: Revising the Agenda of Inquiry and Practice." In R. Paehlke and D. Torgerson, eds., *Managing Leviathan: Environmental Politics and the Administrative State*. London, UK: Bellhaven Press.

Passmore, J. 1974. *Man's Responsibility for Nature*. New York: Scribner's.

Pateman, C. 1970. *Participation and Democratic Theory*. Cambridge, UK: Cambridge University Press.

Pennock, J. R. 1979. *Democratic Political Theory*. Princeton: Princeton University Press.

Perrow, C. 1984. *Normal Accidents*. New York: Basic Books.

Peskin, H. M. 1990. "Alternative Environmental and Resource Accounting Approaches." Paper prepared for the International Conference on the Ecological Economics of Sustainability, Washington, DC, May 21–23.

Peter, K. B. 1990. "Commentary on John S. Dryzek's 'The Environmental Politics of the Good Society.'" Paper presented for the workshop on ecology, Committee on the Political Economy of the Good Society, meeting of the American Political Science Association, San Francisco, August 30 to September 2.

Piasecki, B. W., and Davis, G. A. 1987. *America's Future in Toxic Waste Management: Lessons from Europe*. New York: Quorum Books.

Pierce, J. C. 1989. *Public Knowledge and Environmental Politics in Japan and the United States*. Boulder, CO: Westview Press.

Pierce, J. C.; Lovrich, N. P.; and Matsuoka, M. 1990. "Support for Citizen Participation: A Comparison of American and Japanese Citizens, Activists and Elites." *Western Political Quarterly* 43 (1): 39–59.

Plater, Z. J. B.; Abrams, R. H.; and Goldfarb, W. 1992. *Environmental Law and Policy: A Coursebook on Nature, Law, and Society*. St. Paul, MN: West Publishing.

Portney, K. E. 1984. "Allaying the NIMBY Syndrome: The Potential for Compensation in Hazardous Waste Treatment Facility Siting." *Hazardous Waste* 1 (3): 411–21.

Press, D. 1993. "Environmental Regionalism and the Struggle for California." *Society and Natural Resources*, forthcoming.

Quattrone, G. A., and Tversky, A. 1988. "Contrasting Rational and Psychological Analyses of Political Choice." *American Political Science Review* 82 (3): 719–36.

Rappaport, R. 1992. "Gridlock." *San Francisco Focus*, January.

Repetto, R. 1988. *Public Policies and the Misuse of Forest Resources*. Cambridge, UK: Cambridge University Press.

Repetto, R.; Magrath, W.; Wells, M.; Beer, C.; and Rossini, F. 1989. *Wasting Assets: Natural Resources in the National Accounts*. Washington, DC: World Resources Institute.

Rodman, J. 1980. "Paradigm Change in Political Science: An Ecological Perspective." *American Behavioral Scientist* 24 (1): 49–78.

Rotella, S. 1992. "Mexican Toxic Waste Plant Loses Permit." *Los Angeles Times*, April 3.

Russell, A. M. 1992. "Attorney for Sunnyvale Residents Drops Suit Against Westinghouse." *San Jose Post-Record*, April 15.

Sabatier, P. 1988. "An Advocacy Coalition Framework of Policy Change and the Role of Policy-oriented Learning Therein." *Policy Sciences* 21: 129–68.

———. 1991. "Toward Better Theories of the Policy Process." *PS: Political Science and Politics* 24 (2): 147–56.

Sabatier, P.; Loomis, J.; and McCarthy, C. 1990. "Professional Norms, External Constituencies, and Hierarchical Controls: An Analysis of U.S. Forest Service Planning Decisions." Paper presented at the annual meeting of the American Political Science Association, San Francisco, August 30 to September 2.

Sale, K. 1980. *Human Scale*. New York: Coward, McCann, and Geoghegan.

———. 1991. *Dwellers in the Land: The Bioregional Vision*. Philadelphia: New Society Publishers.

Sax, J. L. 1971. *Defending the Environment: A Strategy for Citizen Action*. New York: Knopf.

Scarce, R. 1990. *Eco-Warriors*. Chicago: Noble Press.

Schmitt, C. 1992. "Money's Gone, but Toxic Sites Remain." *San Jose Mercury News*, March 1.

Schmitt, C.; Carey, P.; and Thurm, S. 1991a. "State Fails to Track Toxic-Waste Cheaters." *San Jose Mercury News*, March 31.

Schmitt, C.; Carey, P.; and Thurm, S. 1991b. "Recycling can be Toxic Roulette." *San Jose Mercury News*, April 1.

Schneider, K. 1992. "Administration Tries to Limit Rule Used to Halt Logging of National Forests." *New York Times*, April 28.

Schumacher, E. F. 1973. *Small Is Beautiful: Economics as if People Mattered*. New York: Harper and Row.

Smith, R. 1985. *Liberalism and American Constitutional Law*. Cambridge, MA: Harvard University Press.

Stone, C. 1974. *Should Trees Have Standing?* Los Altos, CA: William Kaufman.

Taylor, B. P. 1992. *Our Limits Transgressed: Environmental Political Thought in America*. Lawrence, KS: University Press of Kansas.

Thurm, S. 1991. "Toxics Reduction Is a Promise Unkept." *San Jose Mercury News*, April 3.

———. 1992. "Toxics Lurk in Soil, Water." *San Jose Mercury News*, May 4.

Tocqueville, A. de. 1945. *Democracy in America*. New York: Vintage Books.

Trzyna, T. 1992. Personal communication, July.

Twight, B. W.; Lyden, F. J.; and Tuchmann, E. T. 1990. "Constituency Bias in a Federal Career System? A Study of District Rangers of the U.S. Forest Service System." *Administration and Society* 22 (3): 358–89.

U.S. Environmental Protection Agency, region IX. 1989. "Groundwater Contamination at South Bay Superfund Sites: Progress Report." April. San Francisco.

Walker, K. J. 1988. "The Environmental Crisis: A Critique of Neo-Hobbesian Responses." *Polity* 21 (1): 67–81.

Wandesforde-Smith, G. 1990. "Moral Outrage and the Progress of Environmental Policy: What Do We Tell the Next Generation about How to Care for the Earth?" In N. J. Vig and M. Kraft, eds., *Environmental Policy in the 1990s*. Washington, DC: CQ Press.

Warren, M. 1992. "Democracy and Self-Transformation." *American Political Science Review* 86 (1): 8–23.

Weigel, R. H. 1977. "Ideological and Demographic Correlates of Proecology Behavior." *Journal of Social Psychology* 103: 39–47.

Winner, L. 1986. *The Whale and the Reactor: A Search for Limits in an Age of High Technology*. Chicago: University of Chicago Press.

———. 1990. "Response to John Dryzek's 'The Environmental Politics of the Good Society.'" Paper presented for the workshop on ecology, Committee on the Political Economy of the Good Society, annual meeting of the American Political Science Association, San Francisco, August 30 to September 2.

World Commission on Environment and Development. 1987. *Our Common Future*. Oxford, UK: Oxford University Press.

Wright, S. 1991. "The NIMBY Syndrome: Environmental Failure and The Credibility Gap." *HMC* (March/April): 56–58.

Index

About the Author

After receiving an enology degree from the University of California at Davis, Daniel Press worked in the wine industry in California and France. Following on this exposure to science, industry, and agriculture, Press developed a keen interest in environmental affairs, especially at the local and regional levels. Dr. Press is currently Assistant Professor of Environmental Studies at the University of California, Santa Cruz, where he teaches environmental politics and policy, environmental law, and environmental ethics.

Library of Congress Cataloging-in-Publication Data

Press, Daniel, 1962–
Democratic dilemmas in the age of ecology : trees and toxics in the American West / Daniel Press.
p. cm.
Includes bibliographical references and index.
ISBN 0-8223-1503-3. — ISBN 0-8223-1514-9 (pbk.)
1. Environmental policy—West (U.S.)—Citizen participation. 2. Decentralization in government—West (U.S.) 3. Democracy—West (U.S.) I. Title.
GE185.W47P74 1994
363.7'00978—dc20 94-7247 CIP